Ultraviolet Light in Water and Wastewater Sanitation

Ultraviolet Light in Water and Wastewater Sanitation

by
Willy J. Masschelein, Ph.D.

Edited for English by
Rip G. Rice, Ph.D.

CRC Press
Taylor & Francis Group
Boca Raton London New York

CRC Press is an imprint of the
Taylor & Francis Group, an **informa** business

First published 2002 by Lewis Publishers
Taylor & Francis Group
6000 Broken Sound Parkway NW, Suite 300
Boca Raton, FL 33487-2742

Reissued 2018 by CRC Press

© 2002 by Taylor & Francis
CRC Press is an imprint of Taylor & Francis Group, an Informa business

No claim to original U.S. Government works

Publisher's Note
The publisher has gone to great lengths to ensure the quality of this reprint but points out that some imperfections in the original copies may be apparent.

Disclaimer
The publisher has made every effort to trace copyright holders and welcomes correspondence from those they have been unable to contact.

ISBN 13: 978-1-138-10498-3 (hbk)
ISBN 13: 978-1-138-56288-2 (pbk)
ISBN 13: 978-1-315-12134-5 (ebk)

Visit the Taylor & Francis Web site at http://www.taylorandfrancis.com and the CRC Press Web site at http://www.crcpress.com

Preface

Only a few books are available on the subjects of ultraviolet (UV) radiation and its industrial applications in water treatment and sanitation, or on general aspects that may be related to these applications. Refer to books by: [Ellis, 1941; Jagger, 1967; Guillerme, 1974; Kiefer, 1977; Phillips, 1983; Braun, 1986]; and some overview papers exist on UV application to water sanitation; refer to [Gelzhäuser, 1985; Masschelein, 1991, 1996].

A few overview documents also exist [Jepson, 1973; U.S. Department of Commerce, 1979; Scheible, 1985; Gelzhäuser, 1985; Masschelein, 1991, 1996; *J. Water Supply—AQUA*, 1992]. In 1997, the Water Environmental Federation (WEF) published a digest on disinfection in which UV (mainly for wastewater treatment) is reported extensively.

Using animal infectivity as a method of evaluation has indicated that *Cryptosporidium parvum* oocysts may be inactivated significantly by UV irradiation in water treatment. (For further details, see Chapter 3, Table 9.) This finding has thrust UV treatment into the forefront of potable water treatment.

In 1999, the U.S. Environmental Protection Agency (EPA) organized a workshop on UV disinfection of drinking water [U.S. EPA, 1999]. In December 2000, the National Water Research Institute (NWRI), in collaboration with the American Water Works Association Research Foundation (AWWARF) published *Ultraviolet Disinfection Guidelines for Drinking Water and Water Reuse* [NWRI, 2000].

In June 2001, the International Ultraviolet Association (IUVA) held its first International Congress on Ultraviolet Technologies, and the proceedings of that conference [IUVA, 2001] contain many papers on the subject of drinking water disinfection with UV radiation. Furthermore, numerous papers, often also more or less commercially oriented presentations, are available on particular aspects of the application of UV in water treatment.

This text includes discussions of not only disinfection but also removal of recalcitrant micropollutants. On the other hand, no recent monograph is currently available integrating fundamental knowledge, recommendations for design, evaluation of performances, and outlooks for this application. Therefore, the goal of this book is to integrate fundamental knowledge and operational issues.

For some readers who operate systems in the field, certain chapters may be a little lengthy and theoretical. In such cases, I invite them to consult the detailed list of key words in the Glossary. The goal also is to specify the underlying principles of an interesting application that often is still considered as a little empirical in water sanitation practices.

Acknowledgments

The production of the original monograph in French and published by Technique de l'Eau was encouraged by BERSON-UV Technology (the Netherlands). I thank that company for assistance in making an updated monograph available for the purpose of promoting new progress.

The enclosed material is produced privately by the author, who assumes responsibility for the opinions or statements of fact expressed on this emerging technology.

Finally, I am deeply grateful to my long-time and very good friend, Dr. Rip G. Rice, for the editorial assistance with this English version.

Willy J. Masschelein
Brussels, Belgium

Table of Contents

1 Introduction

Critical aspects for the wider application of ultraviolet (UV) light in drinking water treatment sometimes have been described, in spite of the success of the method as established in the field:

- Absence of well-established and generally accepted design rules
- Absence of a permanently active residual agent in the treated water
- Suspicions of the possible photochemical formation of by-products
- Possibility of revival–reactivation by repair mechanisms of irradiated organisms
- Need for operational control of the permanent reliability of the technique

The purpose of this publication is to analyze these concerns and to present extensive information (integrating both fundamental aspects and applications) on the currently available UV technologies applicable to water treatment. These technologies include:

- Lamp technologies available, criteria of evaluation, and choice of technology
- Fundamental principles applicable
- Performance criteria for disinfection
- Design criteria and methods
- Outlook to synergistic use of UV + oxidants
- Functional requirements and potential advantages and drawbacks of the technique

1.1 HISTORICAL: USE OF ULTRAVIOLET LIGHT IN DRINKING WATER TREATMENT

UV radiation can be used for the improvement of drinking water quality. Presently, disinfection is the primary purpose of applying UV irradiation in water treatment. The technical method was introduced by drinking water facilities in the beginning of the twentieth century.

The bactericidal effect of sunlight radiant energy was first reported by Downes and Blunt [1877]. However, the UV part of the sunlight that reaches the earth surface is merely confined to wavelengths higher than 290 nm. The so-called "Boston sunlight on a cloudy day," has a total intensity of 340 W/m^2. However, the instant irradiation that depends on the height of the sun can vary by a factor of 2 to 100. At 30° the total intensity is about 50% higher in high mountains than on flat lands at sea level [Kiefer, 1977]. In addition, only less than 10% of the total sunlight intensity that reaches the surface of the earth is UV light, with little active radiation for water disinfection available from this percentage. Therefore, UV disinfection is essentially a technological process for use in water treatment.

The first large-scale application of UV light, at 200 m^3/day, for drinking water disinfection was in Marseille, France from 1906 to 1909 [Anon., 1910; Clemence, 1911]. This application was followed by a UV disinfection of groundwater for the city of Rouen, France. However, considerable discussions and controversy occurred on the comparative benefit of UV vs. filtration [Anon., 1911]. The applications of UV for water sanitation were delayed in Europe during World War I.

In the United States, the first full-scale application of UV light in 1916 was reported for 12,000 inhabitants of Henderson, Kentucky [Smith, 1917]. Other applications began in Berea, Ohio (1923); Horton, Kansas (1923); and Perrysburg, Ohio (1928). The application of UV in the United States are referenced in early publications of Walden and Powell [1911], von Recklinghausen [1914], Spencer [1917], Fair [1920], and Perkins and Welch [1930].

All these applications were abandoned in the late 1930s. The reasons were unknown but presumably costs, maintenance of the equipment, and aging of the lamps (which at that time, were not fully assessed) were determinants. Disinfection with chlorine probably was preferred for more easy operation and for lower cost at that time. During the 1950s, the UV technique moved into full development again. Kawabata and Harada [1959] reported on necessary disinfecting doses.

In Europe today, over 3000 drinking water facilities use disinfection based on UV irradiation. In Belgium, the first full-scale application was installed and operated in Spontin for the village of Sovet in 1957 and 1958. It is still in operation (see Chapter 3). New applications and technologies are continuously examined and developed. Most of the applications in Europe concerned drinking water or clear water systems, including ultrapure water for pharmaceutical and medical applications. Contrary to those in the United States and Canada, the application to wastewater remained rare, but innovations are under way.

As far as drinking water is concerned, up to 1980 the information on the use of UV in the United States was anecdotal [Malley, 1999]. The EPA Surface Water Treatment Rule (SWTR) of 1989 did not indicate UV as the best available technology for inactivation of *Giardia lamblia*. The proposed Groundwater Disinfection Rule (GWDR) [U.S. EPA, 2000], however, includes UV as a possible technology.

Since 1990, joint research efforts have been made by American Water Works Association (AWWA) and the AWWA Research Foundation (AWWARF). In 1998, it was demonstrated that UV could be appropriate for inactivation of oocysts of *Cryptosporidium parvum*.

In 1986 and 1996, the European Committee of the International Ozone Association organized a symposium [Masschelein, 1986, 1996] on the use of ozone, UV, and also potential synergisms of ozone and UV for water sanitation. The same topics were on the agenda of the IOA Conference at Wasser, Berlin in 2000. At present, the use of these techniques is a major development, perhaps more in the field of wastewater treatment than directly for drinking water, although direct treatment of raw water sources becomes attractive.

Following the developments of ozone–UV, the possibilities of UV in conjunction with hydrogen peroxide and catalysts with UV are actively under examination. Although the applications of these new technologies still remain limited as far as drinking water is concerned, their areas of development include removal of difficult micropollutants (such as herbicides, organochlorine compounds, and polycyclic aromatic hydrocarbons), disinfection, and less formation of by-products.

1.2 PRESENT STATE OF STANDARDS AND REGULATIONS

Only a limited number of official regulations exist for performance-related criteria applicable to UV units for drinking water treatment. At present, in Europe, only Austria officially requires 450 J/m^2 of UV-C irradiation for the disinfection of publicly supplied drinking water [Austria Önorm, 2001].

In Germany, the German Association of Manufacturers of Equipment for Water Treatment (FIGAWA) has published recommendations [FIGAWA, 1987]. Besides giving technical descriptions, these guidelines also recommend applying a minimum appropriate UV dose of 250 J/m^2.

The Deutsches Verein von Gas and Wasserfachmännern (DVGW) has issued recommendations (Arbeitsblatt W 29-4-1997), formulating technical guidelines, particularly concerning the monitoring, and also stipulating a minimum dose of 400 J/m^2. The different recommendations are the basis of some point-of-use applications, for example, for railway trains transporting passengers. Further work is ongoing at DVGW and also at the German Standardization Institute (DIN). It is likely that the German standard will conform with the requirements in Austria.

No DIN standard exists (yet) on the application of UV in water treatment. For general photochemical purposes, refer to the standard DIN-5031-10-1996: Strahlungsphysik im optischen Bereich und Lichttechnik. Other national recommended requirements are Norway, 160 J/m^2; and France, 250 J/m^2. Also the KIWA in the Netherlands has recommended 250 J/m^2 as a minimum dose.

At present no project is under way for a Comité Européen de Normalization (CEN) standard on application of UV in drinking water treatment. However, the issue is under evaluation in different national groups (e.g., DVGW in Germany). An older general recommendation in the United States was an irradiation dose of 240 J/m^2 [Huff, 1965]. Most European countries (including Belgium) rely on this value as a recommendation.

Similar requirements have been formulated for the application of UV disinfection of drinking water onboard ships. In 1966, the U.S. Department of Health, Education and Welfare (DHEW) (now the Department of Health and Human Services) proposed a minimum guideline of 160 J/m^2 for this application, at all points within the disinfection chamber (see also, UK Regulation 29(6) [1973]; and Germany [1973] Vol. 2, Kap.4 [1973]). The application is supposed to be carried out on clear water, pretreated for turbidity and color if required.

The U.S. National Sanitation Foundation (NSF) and American National Standards Institute (ANSI) and NSF Standard 55-1991 define two criteria:

Point-of-use—A dose of 380 J/m^2 is considered safe for disinfection of viruses and bacteria; and 4 log removal of viruses. (The standard also requires that the reactor is validated by the disinfection of challenging bacteria: either *Saccharomyces cerevisiae* or *Bacillus subtilis.*)

Point-of-entry—A dose of 160 J/m^2 is required for supplemental disinfection of municipally treated and disinfected water.

The U.S. EPA SWTR requires a UV dose of 210 and 360 J/m^2 to achieve an abatement of hepatitis virus A (HAV) of 2 and 3 logs, respectively. Most of the states in the United States require compliance with the ANSI and NSF standard requirements mentioned earlier. Exceptions are New Jersey, Pennsylvania, and Utah, which specify a dose of 160 J/m^2. Sometimes a filtration step is required prior to the UV disinfection. AWWA recommends an irradiation dose of 400 J/m^2 for the direct use of UV by small municipal systems.

The Council Directive 91/271/European Economic Community (EEC) concerning urban wastewater treatment does not explicitly require disinfection to be part of the treatment. The requirements are to be defined by local authorities considering the local reuse of the water. Some details are further indicated in Chapter 4.

1.3 DEFINITION OF ULTRAVIOLET LIGHT: RANGE AND NATURAL SOURCES

1.3.1 DEFINITION OF ULTRAVIOLET LIGHT

UV is part of electromagnetic waves. Historically, the nature of light has been the subject of considerable discussion. Newton (1642 to 1727) formulated the corpuscular theory of light, whereas Huyghens (1629 to 1695) promoted the wave theory. The differences in concepts led to considerable analyses in the nineteenth century. The wave theory was supported by the concepts of Maxwell (1831 to 1879), who developed the electromagnetic theory of light, stating that light is composed of electric and magnetic field vectors, orthogonal to each other and constantly traverse the direction of undulary propagation.

The whole discussion came to an end in 1900 when Planck (1858 to 1947) associated and quantified the concept that the photon was associated with the light waves. Newton had shown that white light could be decomposed into visible constituent colors by a prism. However, the visible part of the electromagnetic wave spectrum is only a very small part of the total (Figure 1).

In 1801, using experiments blending sunlight with filters to eliminate the visible components (and also infrared [IR] and higher wavelengths), Ritter was able to show

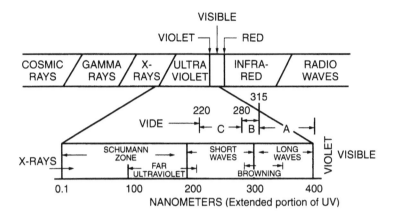

FIGURE 1 Range of electromagnetic waves.

that reduced silver could be produced on irradiation of silver chloride with invisible light of a shorter wavelength than the visible part of the electromagnetic spectrum. One way of blending is by the glass of Wood, which is a glass containing oxides of nickel and cobalt, opaque to visible radiations but transmitting part of the UV rays. This part is called UV-A (Figure 1).

In 1804, Young established the principle of interference with invisible light as detected by paper impregnated with silver chloride. The similar nature of UV light and visible light (interferences of Newton) was therefore established. This also provided an early way of characterizing the wavelengths involved.

1.3.2 ULTRAVIOLET LIGHT RANGES

Ultraviolet light electromagnetic radiation ranges between 400 and 10 nm, and is subdivided into several regions. The range of invisible UV light is established downward from 400-nm wavelengths. As a first stage of evidence, existence could be produced only down to 320 nm, because no optical glass was available to transmit photons of lower wavelengths.

In 1862, Stokes was able to use quartz to extend the perception to 183 nm. From this wavelength downward, oxygen and nitrogen were known to absorb the light. However, Schumann extended the range of observation to 120 nm by using fluorine optics and placing the spectrograph under vacuum. At the beginning of this century, Lyman (1906, 1916, cited in Gladstone, 1955, p. 39) could analyze the solar spectrum down to 5.1 nm when using gratings.

The following classification is more or less empirical but integrates the history of discovery of different UV ranges with the chemical and physiological effects of UV:

Type	Range	Comment
UV-A	From 400 to 315 nm	Between 400 and 300 nm, sometimes called near UV
UV-B	From 315 to 280 nm	Sometimes called medium UV
UV-C	From 280 to 200 nm	Range to be considered in more detail in water disinfection

From 300 to 200 nm the light is also called far UV. From 200 to 185 nm there is some kind of "no man's land" in the definitions. In the vacuum UV light range, several zones are named for their discoverers, as follows:

Range of Schumann, from 185 to 120 nm
Range of Lyman, from 120 to 50 nm
Range of Millikan, from 50 to 10 nm

Under 10 nm, the region of x-rays starts; and at lower wavelengths (under 0.1 nm), the γ-ray region begins.

The whole range of UV light wavelengths is called *actinic waves*, also known as *chemical waves*, in opposition to the *thermic waves* of a higher frequency. Actinic wavelengths involve energies that are able to provoke direct chemical changes in the irradiated molecules (activation, ionization, dissociation, etc.), and to promote biological changes in the systems accordingly.

In the past, the main source of UV light was solar energy. The practical limit to be considered is at wavelengths down to 295 nm, with the atmosphere filtering the lower wavelengths. The maximum effect on skin pigmentation (bronzing) is obtained around 360 nm. Erythema of the skin is promoted by UV, with a maximum effect at about 300 nm and a second maximum around 250 nm (Figure 2).

Different parts of the body, when exposed, may show different sensitivities, but the maximum effects always follow the same pattern. Erythema is a potential disease factor for professionals working with UV and must be prevented by appropriate measures such as spectacles and glass protections.

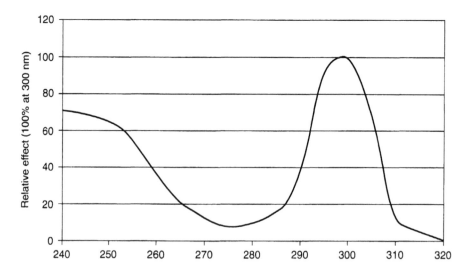

FIGURE 2 Standard erythema curve as defined by the Commission Internationale de l'Eclairage 1939.

1.3.3 DISINFECTION OF WATER WITH ULTRAVIOLET LIGHT

The practice of water disinfection with UV light is mainly concerned with the UV-C range, which means that the optical equipment needs to be as transparent as possible. Quartz remains the best option. Interest is growing in the UV-B range that is able to photolyze proteic and other cellular material (see Chapter 3, Table 7 and Chapter 4).

1.4 SOLAR RADIANT ENERGY

For years, solar radiant energy has been the only known and available source of UV light on the earth. The thermal production of UV light is illustrated in Figure 3, according to the *black body* concept.

With this assumption applied to the solar system, the radiation power is about 4.1023 kW and a corresponding blackbody temperature is estimated at 5780 K. In such conditions (Figure 4), part of solar radiation is in the UV range. The radiant energy received by Earth is estimated at 1400 J/m^2 sec, with the so-called *solar constant* of 1374 W m^{-2}. Most of the emitted light is UV, (about 98%), but only a small part of the emitted UV is received on Earth.

Two basic mechanisms occur: diffusion (scattering) and absorption. The diffusion of Rayleigh is concerned more with short wavelengths because it is proportional to λ^{-4}. Absorption by nitrogen and oxygen eliminates all vacuum ultraviolet (VUV). Wavelengths under 200 nm when absorbed by oxygen, generate ozone, whereas

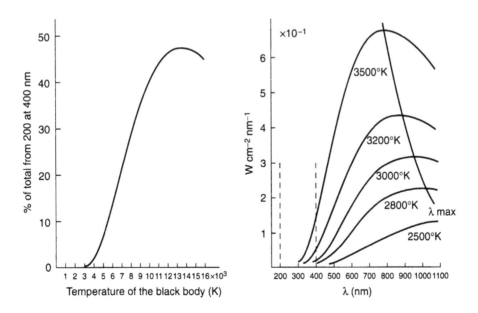

FIGURE 3 Thermal emission of UV (200 to 400 nm) according to the black body theory.

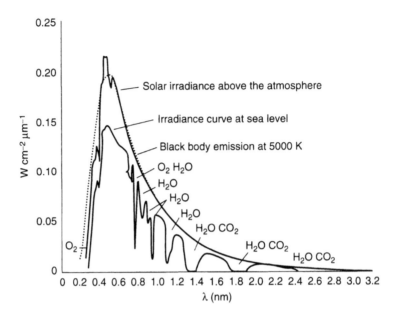

FIGURE 4 Thermal emission curves according to the black body theory.

ozone itself undergoes photolysis when absorbing in the range of 220 to 300 nm. As a consequence, UV-A and a little UV-B are UV components that reach the surface of the Earth. The absorption of UV by ozone, as applicable in water sanitation, is discussed further in Chapter 4.

2 Available Lamp (or Burner) Technologies

2.1 GENERAL

Light can be generated by activating electrons to a higher orbital state of an element; the return of that activated species to lower energy states is accompanied by the emission of light. The process is schematically illustrated in Figure 5.

The quantitative aspects are expressed as $E_1 - E_0 = h\nu$. In other words, wavelengths obtained depend on the energy difference between the activated state and the return state.

Thermal activation of matter provides a means of production of light. According to the black body concept, the *total* radiant power depends on the temperature of the matter and is quantified by the Stefan–Boltzmann law: $P(T) = sT^4$, where $P(T)$ is the total radiant power in watts, radiated into one hemisphere (2π-solid angle) by unit surface at T Kelvin. The Stefan–Boltzmann constant (s) equals 5.6703×10^{-12} W cm^{-2}. However, the emissivity obtained depends on the wavelengths of interest. Black body radiation is not a major source of technological generation of ultraviolet (UV) light, but cannot be entirely neglected in existing lamps either.

2.2 MERCURY EMISSION LAMPS

Activation (or ionization) of mercury atoms by electrons (i.e., electrical discharges) at present is by far the most important technology in generating ultraviolet (UV) light as applicable to water disinfection. The reasons for the prevalence of mercury are that it is the most volatile metal element for which activation in the gas phase can be obtained at temperatures compatible with the structures of the lamps. Moreover, it has an ionization energy low enough to enable the so-called "avalanche effect," which is a chain reaction underlying the electrical discharge. A vapor pressure diagram is given in Figure 6.

Activation–ionization by collision with electrons and return to a lower energy state (e.g., the ground state) is the principle of production of light in the system (see Figure 5).

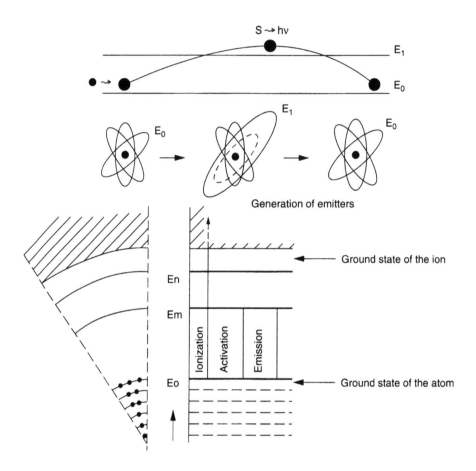

FIGURE 5 Emission of radiation by matter (schematic).

As for the energy diagram or Grothian diagram for mercury, refer to Figure 7. As a first conclusion, there is a whole series of return levels from the ionized or the activated metastable states appropriate for emitting in the UV range.

Natural mercury is composed of five isotopes at approximately equal weight proportions; thus small differences in the line emissions exist, particularly at higher vapor pressures, and give band spectra instead of line emissions.

2.2.1 EFFECT OF FILLER GAS: PENNING MIXTURES

The most used filler gas is argon, followed by other inert gases. These gases have completed outer electron shells and high ionization energies as indicated in Table 1.

In most technologies, argon is used as filler gas. The ionization energy of argon is 15.8 eV, but the lowest activated metastable state is at 11.6 eV. The energy of this metastable state can be lost by collision. If it is by collision with a mercury atom, ionization of the latter can take place and this can be followed by emission of light. When the energy of the metastable state is higher than the ionization energy of

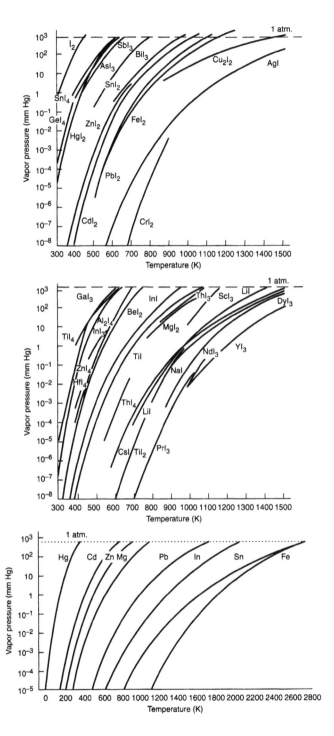

FIGURE 6 Vapor pressure diagram of elements and compounds of interest in the generation of UV light.

TABLE 1
Ionization Energies of Inert Gases vs Mercury (Values in eV)

Element	Ionization Energy	Energy of Lowest Excited State
Mercury	10.4	4.77
Xenon	12.1	8.32
Krypton	14.0	9.91
Argon	15.8	11.6
Neon	21.6	16.6
Helium	24.6	19.8

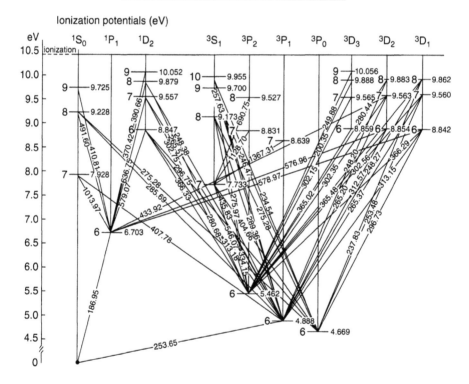

FIGURE 7 Grothian diagram of the mercury atom.

mercury, the whole constitutes a Penning mixture. Consequently, Penning mixtures are possible with mercury, argon, neon, helium, but not with krypton and xenon.

The primary role of the filling gases is not only to facilitate the starting of the discharge but also to promote the starting activation–ionization of the mercury. The filler gas is usually in excess of gaseous mercury; however, if the excess is too high, energy of the electrons can be lost by elastic collisions with filler gas atoms, thus decreasing the emission yields by thermal losses.

2.3 CURRENTLY AVAILABLE COMMERCIAL LAMP TECHNOLOGIES

2.3.1 LOW-PRESSURE MERCURY LAMP TECHNOLOGIES

Mercury lamps are operated at different mercury-gas pressures. The low-pressure mercury lamp for the generation of UV normally is operated at a nominal *total gas pressure* in the range of 10^2 to 10^3 Pa (0.01 to 0.001 mbar), the carrier gas is in excess in a proportion of 10 to 100. In low-pressure Hg lamps, liquid mercury always remains present in excess at the thermic equilibrium conditions installed.

2.3.2 MEDIUM-PRESSURE LAMP TECHNOLOGIES

The medium-pressure mercury lamp operates at a total gas pressure range of 10 to 30 MPa (1 to 3 bar). Normally, in medium-pressure mercury lamps, no liquid mercury is permanently present in excess at nominal operating conditions.

Both lamps are based on plasma emission at an inside lamp temperature of 5000 to 7000 K; in the low-pressure technology the electron temperature must be high, whereas in the medium pressure technology electron and atom ion temperature comes to equilibrium (Figure 8). Depending on the exact composition of the gas mixture, and the presence of trace elements, and the electrical feed parameters, the emission in the UV range of medium-pressure Hg lamps can be modified into, for example, broadband emission or multiwave emission (further details in Section 2.4.2.3).

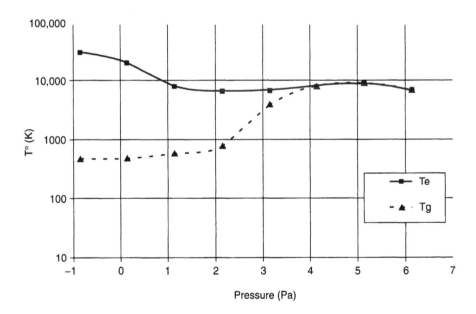

FIGURE 8 Plasma temperatures in mercury discharge lamps (schematic) (T_e and T_g, temperature of electrons and of the gas phase, respectively).

2.3.3 High-Pressure Mercury Lamps

High-pressure mercury lamps are used less in water treatment. Such lamps operate at pressures (total), up to 10^6 Pa (10 atm), emitting continuous spectra less appropriate for specific applications like water disinfection or specific photochemical reactions.

2.4 AVAILABLE LAMP TECHNOLOGIES

The next sections specifically report on the low- and medium-pressure mercury lamps and secondarily on special lamp technologies. Flash-output lamps and excimer lamps are interesting developments, but no significant applications have been found yet for large-scale water treatment.

Note: Some confusion exists in the literature in the pressure terminology of UV lamps. In actinic applications, a field to which water treatment also belongs, the classification is low-pressure; medium-pressure, and eventually high-pressure. When illumination is concerned, one finds low-pressure, high-pressure, and less termed as very high-pressure as corresponding labels. That is why in the practical field of application in water treatment, medium-pressure and high-pressure mercury lamps correspond to the same concept.

2.4.1 Low-Pressure Mercury Lamp Technologies

2.4.1.1 General Principles

In low-pressure technology, the partial pressure of mercury inside the lamp is about 1 Pa (10^{-5} atm). This corresponds to the vapor pressure of liquid mercury at an optimum temperature of 40°C at the lamp wall. The most simple way to represent the process of generation is to consider the ionization of atomic mercury by transfer of kinetic energy from electrons upon inelastic collisions with the mercury atoms:

$$Hg + e = 2e + Hg^+$$

In theory, the proportion of ionized mercury atoms is proportional to the electron density in the discharge current. However, electron–ion recombinations can occur as well, thus reconstituting the atomic mercury. The whole of the ionization process involves a series of steps in which the Penning effect of the filler gas is important, particularly during the starting or ignition period of the lamp:

$$e + Ar = Ar^*(+e)$$

$$Ar^*(+e) + Hg = Hg^+ + e + Ar$$

At a permanent regime of discharge, the electrons in the low-pressure mercury plasma do not have enough kinetic energy to provoke direct ionization in one single step, and several collisions are necessary with formation of intermediate excited mercury atoms:

$$e + Hg = Hg^*(e)$$

$$Hg^*(e) + e = 2e + Hg^+$$

The reaction by which a photon is emitted corresponds to:

$$Hg^* \text{ (excited state)} \rightarrow Hg \text{ (ground state)} + h\nu$$

or

$$Hg^* \text{ (excited state)} \rightarrow Hg^* \text{ (less excited state)} + h\nu$$

The permissible quanta are those indicated in the Grothian diagram for mercury (see Figure 7). The emission of a photon by an atom in an excited electronic state is reversible; this means that before escaping from the plasma contained in the lamp enclosure the emitted photons can be reabsorbed by another mercury atom. This phenomenon is called *self-absorption*, and becomes naturally more important when the concentration of ions in the gas phase is increased and the pathway of the photons is longer (higher lamp diameters). For mercury lamps, self-absorption is most important for the 185- and 253.7-nm lines. Overall, the reversibility in emission–absorption is translated in the low-Hg pressure technology, by a higher emission rate near the walls of the lamp than from the inside parts of the plasma.

Low-pressure mercury lamps usually are cylindrical (with the exception of the flat lamp technology; see Section 2.5.1). They are currently available in lamp diameter ranges from 0.9 to 4 cm, and lengths of 10 to 160 cm. Along the length of a tubular discharge lamp the electrical field is not uniform, and several zones can be distinguished (Figure 9).

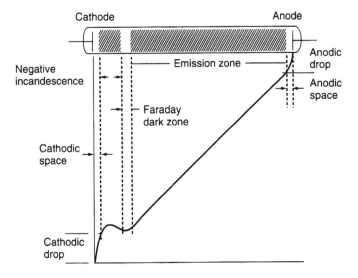

FIGURE 9 Discharge zones in a tubular lamp.

Besides the drop-off of emitted intensity at the cathode, on the cathode side there is a Faraday dark space of about 1-cm length. The dark spaces at constant lamp pressure remain constant, whereas the emissive range expands according to the total length of the lamp. This means that for short lamps the useful emission length is proportionally shorter than for long lamps. To account for this phenomenon, the manufacturers constructed U-shaped and other bent lamps (examples in Figure 10) to meet the geometric conditions in the case of need for short low-pressure Hg lamps.

2.4.1.2 Electrical Feed System

In practice, the low-pressure mercury lamps are supplied by alternative current sources, with the cathode and anode sides constantly alternating, as will the Faraday dark space. Moreover, the ionization generates an electron-ion pair of a lifetime of about 1 msec. However, on voltage drop, the electrons lose their kinetic energy within microseconds. As the lamps are operated with moderate frequencies, at the inversion point of the current half-cycles, the emission is practically extinguished. This is in contrast with medium-pressure technologies.

The electrical current feed can be of the cold, or of the hot cathode type. The cold cathode type is a massive construction with electrodes (generally) in iron or nickel that needs bombardment of the cathode by positive ions to release electrons into the plasma. This implies that high starting voltages are necessary (up to 2 kV), which are not directly supplied by the mains. The cold cathode type is less applied in water treatment.

The hot cathode type is based on thermoionic emission of electrons from a structured electrode system composed of coiled tungsten wires coated and embedded with alkaline earth oxides: CaO, BaO, or SrO. On heating, the oxide coatings build

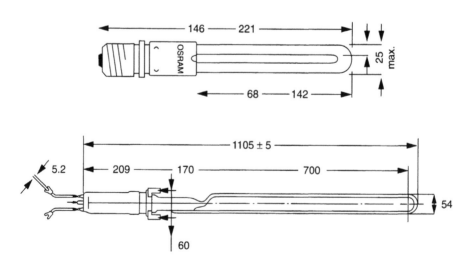

FIGURE 10 U-shaped and bent low-pressure mercury lamps. (Typical sizes given are in millimeters, depending on the manufacturer.)

up a layer of metal (e.g., barium) and at about 800°C enough electrons are discharged to get the emission started. At normal operation regime, the temperatures of the electrodes reach 2000°C. Hot cathode lamps operate at low voltage ranges, (e.g., with voltages of the mains [220 V in Europe]). The cathode possibly can be brought to the necessary discharge temperature in a way similar to that of fluorescent lighting lamps. A typical example of the electrical feed scheme of the hot cathode lamp type is shown in Figure 11.

2.4.1.3 Factors Influencing Emitted Intensity

2.4.1.3.1 Voltage
The effect of fluctuations in voltage of the supply by the mains have a direct influence on the UV output yield of low-pressure mercury lamps (Figure 12).

2.4.1.3.2 Temperature
Temperature outside the lamp has a direct influence on the output yield (Figure 13). Temperature only has a marginal effect by itself, but directly influences the equilibrium vapor pressure of the mercury along the inner wall of the lamp. If too low, the Hg vapor is cooled and partially condensed and the emission yield drops. If too hot, the mercury pressure is increased, as long as there is excess of liquid Hg. However, self-absorption is increased accordingly and the emission yield is dropped. The optimum pressure of mercury is about 1 Pa, and the optimum temperature is around 40°C.

Curve 1 in Figure 13 is for lamps in contact with air and curve 2 with water; both are at temperatures as indicated in the abscissa. They are in line with the differences in heat capacities between air and water.

An important conclusion for water treatment practice is that the lamps should be *mounted within a quartz tube* preferably with open ends through which air is circulating freely to moderate the effects of cooling by water. This is more important when cold groundwater is treated. The effect of temperature can be moderated by using amalgams associated or not associated with halides (see later the flat lamp indium-doped technology and the SbI_3-A lamp technology).

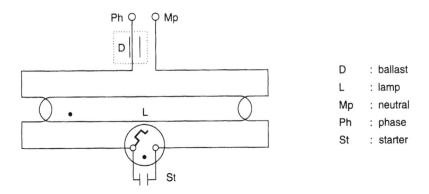

D	: ballast
L	: lamp
Mp	: neutral
Ph	: phase
St	: starter

FIGURE 11 Typical electrical feed system of a low-pressure Hg lamp.

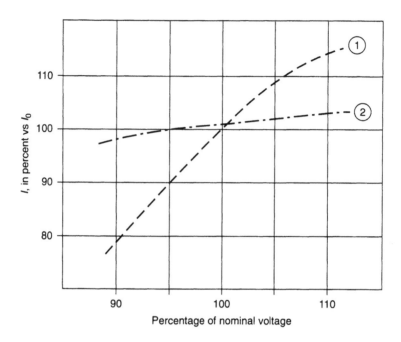

FIGURE 12 Influence of voltage (of off-take from the mains vs. nominal) of supply current on UV output. (Curve 1 is for low-pressure lamps; curve 2 is for medium-pressure lamps.)

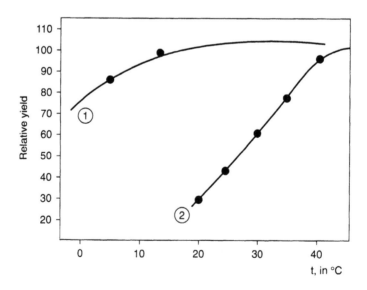

FIGURE 13 Temperature effect on 254-nm radiation of a typical low-pressure germicidal Hg lamp.

2.4.1.3.3 Aging of Lamps

Figure 14 gives a typical example of aging characteristics of low-pressure Hg lamps. During the first 100 to 200 h of operation an initial drop in emission yield occurs. After that period the emission is stable for months.

The main cause of aging is solarization of the lamp wall material (the phenomenon is faster for optical glass than for quartz); the secondary cause is by blackening due to deposits of sputtered oxides from the electrodes. Under normal conditions, low-pressure Hg lamps are fully operational for at least 1 year.

Note: One start–stop procedure determines an aging rate equivalent to that of 1-h nominal operation.

For aging of low-pressure mercury lamps that emit for photochemical oxidation processes at 185 nm, see Chapter 4.

2.4.1.4 Typical Emission Spectrum

The most usual low-pressure mercury lamp emission spectrum is illustrated in Figure 15. The spectrum is of the line or ray type; the emission is concentrated at a limited number of well-defined lines and the source is called monochromatic. The resonance lines at 253.7 and 185 nm are by far the most important. The lines in the 300-nm range and higher can be neglected in water treatment (they can be slightly increased if the pressure of the mercury vapor is increased). The 253.7-nm line represents around 85% of the total UV intensity emitted and is directly useful for disinfection.

The 185-nm line is not directly useful in disinfection and is best eliminated, because by dissociation of molecular oxygen it can eventually promote side reactions with

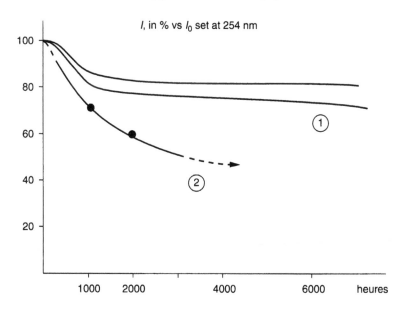

FIGURE 14 Drop in emission yield on aging (at 254 nm). 1 is for conventional low-pressure Hg germicidal lamps; 2 is for indium-doped lamps (1992 technology).

organic components of the water. This elimination can be achieved by using appropriate lamp materials such as optical glass or quartz doped with titanium dioxide.

The relative emission of intensity vs. the most important line at 254 nm (quoted
as 100%) is in the range shown in Table 2 for conventional low-pressure Hg lamps
(i.e., the so-called germicidal lamps according to Calvert and Pitts [1966]).

2.4.1.5 Photochemical Yield

The specific electrical loading in the glow zone, expressed in watts per centimeters,
typically is between 0.4 and 0.6 W(e)/cm. The linear total UV output of the discharge
length for lamps appropriate for use in disinfection is in the range of 0.2 to 0.3 W(UV)/cm.

TABLE 2
Emitted Intensities of Low-Pressure Hg Lamps

λ (nm)	Emitted Intensity (I_o, rel)	λ (nm)	Emitted Intensity (I_o, rel)
184.9	8	289.4	0.04
296.7	0.2	405.5–407.8	0.39
248.2	0.01	302.2–302.8	0.06
253.7	(100)	312.6–313.2	0.6
265.(2–5)	0.05	334.1	0.03
275.3	0.03	365.0–366.3	0.54
280.4	0.02		

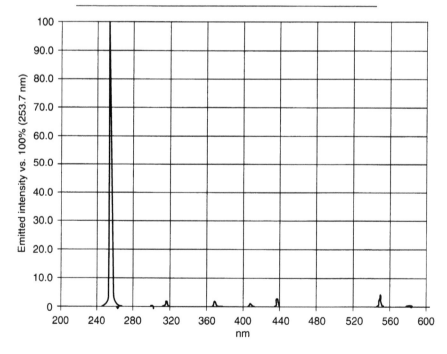

FIGURE 15 Emission spectrum of low-pressure Hg lamps (germicidal lamps).

This means that the UV efficiency generally designed by total W(UV) output vs. W(e) input is between 0.25 and 0.45. The energy losses are mainly in the form of heat (about 90% of them), and emission in the visible (and infrared [IR]) range.

Note: Glow discharge mercury lamps (Figure 16) need a high specific electrical loading, up to 0.85 W/cm; and have a low linear output, in the range of 0.01 to 0.015 W(UV)/cm, with a UV efficiency of about 1.5%. This type of UV source has not been designed for water treatment but is easy for use in experiments in the laboratory [Masschelein et al., 1989].

For low-pressure Hg lamps, the overall UV-C proportion of the UV light wavelengths emitted are in the range of 80 to 90% of the *total* UV power as emitted. These data determine the ratio of useful UV light in disinfection vs. the lamp emission capabilities (see also Chapter 3).

Increasing the linear (UV-C) output is a challenge for upgrading the low-pressure Hg lamp technologies as applicable to water treatment to reduce the number of lamps to be installed. By cooling part of the lamp, it is possible to maintain a low pressure of gaseous mercury (i.e., the equilibrium pressure at the optimum 40°C) even at higher lamp temperatures and hence at higher current discharge.

Designs [Phillips, 1983, p. 200] are based on narrow tubes to reduce the self-absorption and using neon-containing traces (less than 1% of the total gas pressure) of argon at 300 Pa as Penning mixture. The gas is cooled behind the electrodes in cooling chambers [Sadoski and Roche, 1976].

In another design (Figure 17), the UV yield is increased further by constructing long lamps, from 1 to 4 m. The tubes are of the bend type to reduce the necessary space for installation in treatment of large water flows: 75 to 150 m^3 per unit. The specific electrical loading can range from 10 to 30 W/cm glow zone. The UV-efficiency

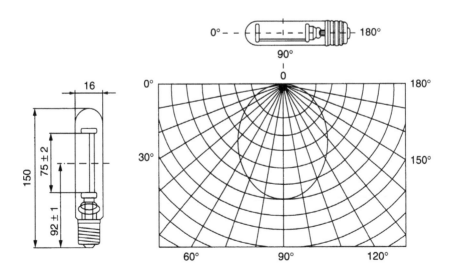

FIGURE 16 Glow discharge Hg lamp from Philips, available in 4, 6, and 8 W(e).

FIGURE 17 Small diameter, multibend-type, high-intensity, low-pressure Hg lamps (formerly BBC).

range is $\eta = 0.3$ with about 90% emission at 253.7 nm. By considering the higher temperature in the discharge zone of 100 to 200°C and the higher radiation density, the high-yield lamp is subject to faster aging than the conventional constructions. An efficient lifetime of 4000 h is presently obtained and the manufacturers are making efforts to improve the lifetime.

See Section 2.5.1 and Figures 24 and 25 for another technology.

2.4.2 MEDIUM- AND HIGH-PRESSURE MERCURY LAMP TECHNOLOGIES

2.4.2.1 General

The medium-pressure mercury lamps operate at a total gas pressure in the range of 10^4 to 10^6 Pa. At nominal operating temperatures of 6000 K in the discharge arc (possible range is 5000 to 7000 K), all the mercury within the lamp enclosure is gaseous. Consequently, the precise amount of mercury to be introduced in the lamps is one of the challenges for manufacturers.

The entire compromise between electron temperature and gas temperature for mercury lamps is illustrated in Figure 8. It can be stated that the coolest possible part of a medium-pressure mercury lamp by the present state of technology is about

400°C, whereas in a stable operation the temperature in the main body of the lamp is in the range of 600 to 800°C.

These operating temperatures make the use of an open (possibly vented), quartz enclosure of the lamp absolutely necessary to avoid direct contact of the surface of the lamp with water. The total heat loss of the lamp is given by the Waymouth formula [Waymouth, 1971]:

$$H = 4\, kp(To - Tw)$$

where

To and Tw = absolute temperatures, in the center and at the wall of the lamp, respectively

k = the thermal conductivity of mercury

H ranges from 9 to 10 W/cm.

Because the center of the lamp is at about 6000 K and the wall is at 1000 K, there is a radial temperature distribution. This distribution is of the parabolic type, $F(r^2)$, with lowered distribution starting from the central axis of the lamp. The true emissive part of the plasma can be considered as located at about two-thirds of the outside diameter of the lamp.

The precise mercury dosing is given by the Elenbaas [1951], equation, which experimentally correlates the mercury vapor pressure (developed at nominal regime) to the mass (m) of mercury enclosed (in milligrams per centimeter arc length) as a function of the diameter of the lamp (d, in cm):

$$P \text{ (in pascal)} = (1.3 \times 10^5 \times m)/d^2$$

The effective mercury pressure in the discharge zone mostly is in the range of 40×10^3 Pa.

Relations also have been formulated [Lowke and Zollweg, 1975] to correlate the mean potential gradient (in volts per centimeter arc length) as a function of the wattage and mercury fill:

$$E \text{ (volt/cm)} = [(P^{1/2})/(P - 4.5 \times P^{1/4})^{1/3}] \times m^{7/12} \times d^{-3/2}$$

wherein P is $(\text{Watt})^{1/6} \times m^{7/12} \times cm^{-9/4}$.

Medium-pressure lamps operate in the potential gradient range of 5 to 30 V/cm. By considering a warm-up value of 20 W/cm, from the preceding relation a quantity of evaporated mercury of about 1 mg/cm arc length is found. Total quantity enclosed is 5 to 10 mg/cm.

2.4.2.2 Emission of UV Light

The emission of medium-pressure mercury lamps is polychromatic (Figure 18) and results from a series of emissions in the UV region and in the visible and IR range as well (Table 3).

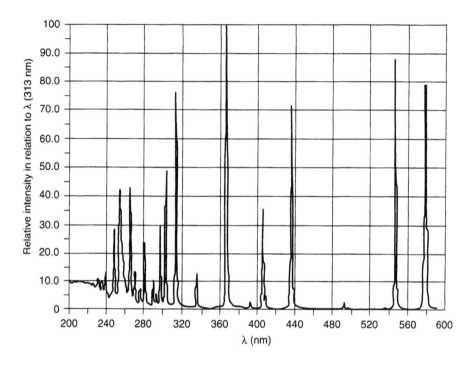

FIGURE 18 Typical emission spectrum of a medium-pressure Hg lamp (100% emission defined at 313 nm).

Note: To optimize the emission in the UV-C range, and consequently the reaction and disinfection capabilities, broadband and multiwave medium-pressure lamps have been developed by Berson. An example of emission in this technology is indicated later in Figure 22.

One can also observe a continuum of emission at 200 to 240 nm. This is usually cut off by the lamp wall material, except if used in the application.

Elenbaas [1951] has measured the total radiant power emitted as a function of the electrical power input and proposed two correlations:

$$P(rad) = 0.72(Pe - 10)$$

and:

$$P(rad) = 0.75(Pe - 4.5\ Pe^{(1/4)})$$

The relations confirm the total intensity of irradiance yield of 65%. However, only part of the intensity is in the specific UV range necessary and potentially useful for disinfection.

2.4.2.3 Voltage Input vs. UV Output

The electrode structure and materials of medium-pressure Hg lamps must meet severe conditions. The temperature of the cathodes is about 2000°C. The thickness of the vitreous silica walls is 1 to 2 mm. A schematic diagram of a medium-pressure UV lamp is given in Figure 19.

TABLE 3
Main Spectral Bands Emitted by a Medium-Pressure Hg Lamp

λ (nm)	Hg Activated State (eV)	Relative Intensity		
		a	b	c
248.3	9.879–9.882	46	28	21
253–260	4.888	5	43	32
265.3	9.557–9.560	10	43	32
269.9	10.056	10	12	9
280.3	9.888	10	24	18
296.7	8.847	20 ·	30	23
302.3	9.560–9.565	40	48	36
313	8.847–8.854	100	75	56
365	8.847–8.859	71–90	100	75
404.7	7.733	39	36	27
407.8	7.928	6	8	6
435.8	7.733	68	71	53
546.1	7.733	80	88	65
577	8.854	82	—	—
579	8.847	83	78	59

Note: Transitions according the Grothius diagram.

[a] Setting 100% at the 313-nm line (typical lamp Philips HTQ-14); 100% corresponds to 200 W (UV) output in a 5-nm range 310 to 315 nm.

[b] Setting 100% at the 365-nm line (Original-Hanau Mitteldruckstrahler). (In this technology a continuum emission of about 10% vs. the 365-nm line exists in the range of 200 to 240 nm.)

[c] For comparison of the yield of [b] vs. the earlier reference [a], one must apply a correction factor of 0.75.

The UV output is approximately directly proportional to the input voltage that also determines (the high voltage) the average power input to the lamps. The correlation holds between 160 and 250 V (voltage of the mains). The precise correlation, I vs. W(e), also depends on the ballast and the transformer, but it is important to note that for a given condition of the hardware, the correlation is about linear.

Small lamps (i.e., up to 4 kW) can be operated on regime by connection to the main current of 220/380 V. A pulse start is necessary with pulse at 3 to 5 kV. For higher lamp power, a high potential transformer is necessary. The latter is recommended anyway, because it is a method of automatically monitoring the lamp output. On increasing the lamp feed high potential, the UV output is increased accordingly (Figure 20).

In addition, the lamp material must have a low thermal expansion coefficient $(5 \times 10^{-7}$ per Kelvin). In present technologies, electrode connections consist of thin sheets of molybdenum (thickness less than 75 μm; thermal expansion coefficient

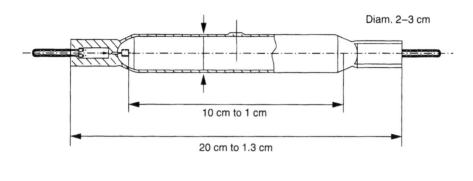

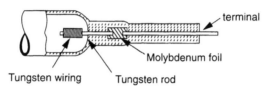

FIGURE 19 General construction of a medium-pressure Hg lamp (example).

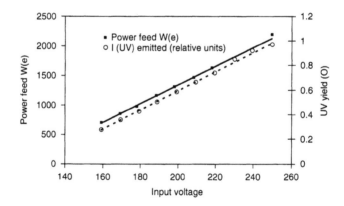

FIGURE 20 Correlation of input voltage (and power input) and UV output of medium-pressure Hg lamps (example).

5×10^{-6} per Kelvin), sealed in the quartz ends and connected inside the lamp to a tungsten rod surrounded by a tungsten wiring. At nominal operating conditions, cathode temperature ranges between 350 and 400°C, but at the tips, temperatures are between 1500 and 2000°C.

The normal (i.e., nominal) thermoionic emission from a cathode is given by the equation:

$$J = A \, T^2 \exp - f(e/kT)$$

where

J	= current density (ampere per square centimeter)
T	= Kelvin temperature
e	= charge of an electron (1.6×10^{-19} C)
k	= the Boltzmann constant (1.372×10^{23} J/K)
A	= emission coefficient of the electrode material, which for pure metals is in the range of 120 A-cm^{-2} K^{-2}
f (in eV)	= practical work function correlating the thermoionic emission rate for a given electrode surface. Values for f are 4.5 eV for tungsten. To reduce this high value, oxide-coating is made between the windings of the electrode wires with alkaline earth oxides or thorium oxide. During operation, the oxide is reduced by tungsten conducting to the formation of the native metal [Waymouth, 1971], which moves to the ends of the electrode rod. The work function is diminished accordingly to 3.4 eV for pure thorium, and 2.1 eV for pure barium. However, monolayers of barium on tungsten have a work function of 1.56 eV and thorium on tungsten of 2.63 eV [Smithells, 1976]. This makes the emission coefficients for Ba/W and Th/W ranges 1.5 and 3.0 A cm^{-2} K^{-2}, respectively. These coefficients enable favorable electrical start conditions of the lamps.

On increasing the high voltage (also the power) increased intensity is emitted and monitoring and automation are possible. However, broadening of the spectral bands occurs simultaneously and must be accounted for appropriately. The overall compromise can be computer-controlled. A typical example of a broadened UV emission spectrum is given in Figure 21.

On start-up, the lamp emits UV light of the same type as the low-pressure Hg lamp with predominantly the resonance lines at 185 and 253.7 nm. The emission gradually evolves to the polychromatic type as illustrated in Figures 18 to 22(a) and (b).

Figure 23 shows examples of Berson medium-pressure lamps.

Overall, in the medium-pressure technologies, the continuum around 220 nm (sometimes called *molecular radiation*) probably is due to braking effects (Bremstrahlung) by collisions of atoms and electrons. The importance of this continuum is related to the square of mercury pressure, and its shape also depends on mercury pressure. If the goal is disinfection and not photochemical oxidation, the entire range under 220 nm can be cut off by the material of the lamp enclosure.

2.4.2.4   Aging

A classical lifetime to maintain at least 80% of emission of germicidal wavelengths is generally 4000 h of operation. In recent technologies, lifetimes from 8,000 to 10,000 h have been reached. Also important is that with aging, the spectrum is modified. Figure 24 gives an indication of the relative output of aged and new lamps at different wavelengths of interest.

In the most recent developments, optimization of the electrical parameter enables the production of lamps emitting up to 30% of the light in the UV-C range. These lamps are operated at an electrical load of 120 to 180 W/cm.

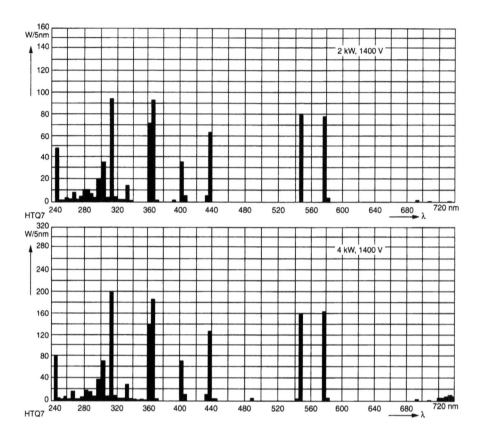

FIGURE 21 Enhanced emissions on increase of power input to medium-pressure Hg lamps. (From documents of Philips, Eindhoven, the Netherlands.)

2.5   SPECIAL LAMP TECHNOLOGIES

2.5.1   FLAT LAMP TECHNOLOGIES

Theoretical aspects related to emission from noncircular lamps were formulated earlier [Cayless, 1960]. The Power Groove lamp (from General Electric) is a flattened U-shaped lamp that was claimed to give higher output than comparable circular lamps [Aicher and Lemmers, 1957].

A flat lamp technology is marketed by Heraeus, Hanau, Germany. This particular technology of low-pressure Hg lamps is based on the construction of lamps with a flat cross-section (ratio of long to short axes of the ellipse of 2:1, Figure 25). This design increases the external surface compared with the cylindrical construction. The ambient cooling is improved accordingly.

For a given gas volume, the travel distance of the photon inside the lamp is less than in an equal cylindrical volume, and the probability of reabsorption is reduced accordingly. The spectral distribution is different (see also Figure 24 for clarification).

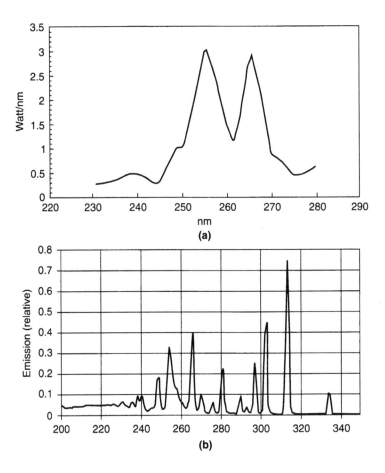

FIGURE 22 (a) Emission of a medium-pressure broadband Hg lamp. (From documents of Berson Milieutechniek, Neunen, the Netherlands.) (b) Emission of the recent Berson multiwave, high-intensity, medium-pressure lamps. (To be considered: the relatively low emission at 220 nm and lower, and a contribution in the range of 300 to 320 nm.)

For more comments on the importance of these components, see Section 3.2.3, but the lifetime is the same as for conventional lamps. The emission at the flat side is about three times higher than at the small side. The technology exists with conventional low-pressure Hg filling, but also in a thermal execution as Spectratherm[R] (registered trade name), in which the mercury is doped with indium. This lamp is also constructed with cooling spots that make operation at higher plasma temperatures possible. This thermal variant can operate at nearly constant emission yield in direct contact with water in the range of temperatures from 10 to 70°C. This makes the construction also appropriate for treatment of air-conditioning and bathing water, as well as for drinking water treatment.

Cylindrical constructions, more easy to manufacture, can emit overall the same intensity. In the flat lamp technology, the relatively higher intensity emitted at the flat side implicates a lower emission at the curved side.

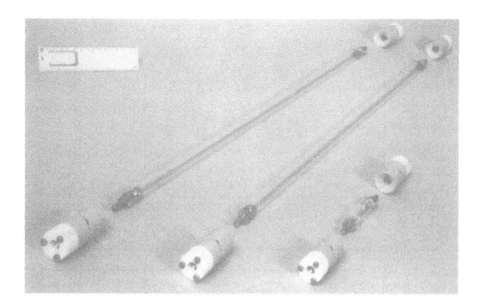

FIGURE 23 Photograph of typical Berson lamps.

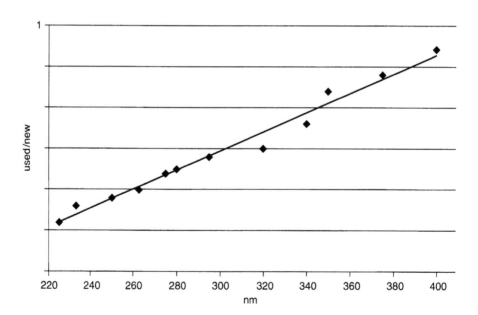

FIGURE 24 Spectral changes on aging (4000 h of continuous operation) of medium-pressure Hg lamps.

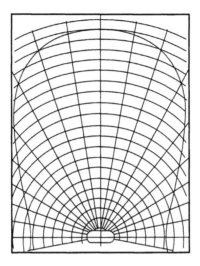

FIGURE 25 Schematic of the zonal distribution of a UV flat-shaped lamp. (Egberts, 1989.)

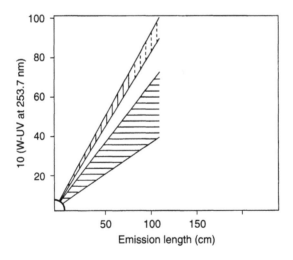

FIGURE 26 Emission of flat-type, low-pressure Hg lamps. (Egberts, 1989.)

Up to now, the flat lamps have been constructed with a maximum length of 112 cm. Total UV light emitted by the flat lamps ranges from 0.6 to 0.7 W (UV)/cm arc length. Comparison is given in Figure 26. The same overall yield also can be obtained with cylindrical lamps.

2.5.2 INDIUM- AND YTTRIUM-DOPED LAMPS

One of the difficulties in design and operation of low-pressure Hg lamp reactors is the temperature dependence of the intensity emitted (see Figure 13). To obviate this problem, doped lamps have been developed. By doping the Penning gas with indium, a more constant emission can be obtained (Figure 27). Also, the doping of Hg lamps can be achieved in the form of amalgams. Yttrium-doped lamps (by Philips) were proposed by Altena (2001). These lamps have similar performances independent of temperature as the Spectratherm lamp.

2.5.3 CARRIER GAS DOPED LAMPS

By modifying the composition of the Penning gas, the output yield can be modified and sometimes improved, but also the spectrum of the emitted light can be changed. Neon has a higher electron diffusion capability than does argon. Incorporating neon together with argon in the Penning mixture provides easier starting and can produce increased linear output [Shadoski and Roche, 1976]. Condensation chambers located behind the electrodes are necessary to maintain the optimum mercury pressure.

2.5.3.1 Xenon Discharge Lamps

Xenon discharge lamps in the medium-pressure range (to high-pressure, i.e., on the order of 10 kPa), emit a spectrum, similar to that of solar radiation (Figure 28).

An available technology that also emits significantly in the 240- to 200-nm range is produced by Heraeus, Hanau, Germany, based on a xenon-modified Penning mixture. The spectral distribution is indicated in Figure 29.

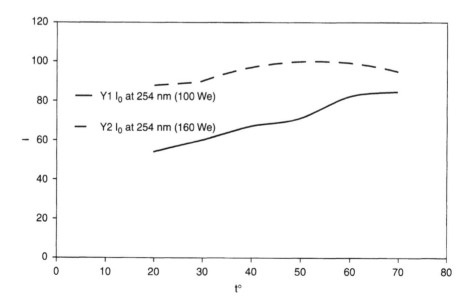

FIGURE 27 Emission of indium-doped lamps at 253.7 nm. (Egberts, 1989, for the Spektratherm™ lamp.) (Spektratherm is a registered trademark from Heraeus, Hanau, Germany; commercial variants exist.)

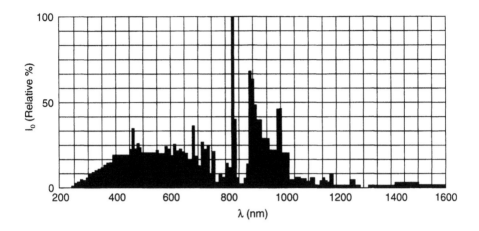

FIGURE 28 Relative light power distribution of xenon discharge lamps. (According to documents of Philips, Eindhoven, the Netherlands.)

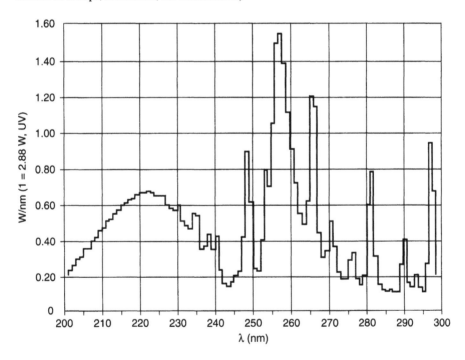

FIGURE 29 Spectral distribution of Xenon-doped, low-pressure Hg lamps. (From documents of Heraeus, Hanau, Germany.)

2.5.3.2 Deuterium Carrier Gas Discharge

Deuterium carrier gas discharge (medium- to high-pressure) lamps have increased emission in the UV-C range, particularly below 250 nm (Figure 30). Lamps based on discharges in carrier gases have not yet been found useful in water treatment,

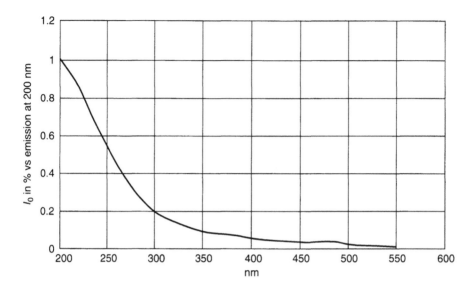

FIGURE 30 Typical emission of a deuterium discharge lamp.

considering the operating costs involved in the treatment of large water volumes or high flows.

2.5.3.3 Metal Halide Lamps

The addition of metal iodides to the Penning mixture changes the spectral distribution. The technique is applied mostly in medium-high-pressure lamps and the spectra are polychromatic with the following dominant ranges: SbI_3, 207 to 327 nm; CoI_2, 345 to 353 nm; FeI_2, 372 to 440; GaI_3, 403 to 417 nm; MgI_2, 285 to 384 nm; PbI_2, 368 to 406 nm; and TlI, 535 nm.

Among the halide-doped lamps in water treatment practice, essentially the antimony iodide doping elicits interest. In some variants, no mercury is used in the filler gas, but only xenon or neon [Schäfer, 1979]. The lamps are operated at 10^5 Pa total pressure, generating (total) UV power with a yield of 12 to 15%, at a high linear output of 3.5 to 4.5 W/cm glow zone. This high linear output enables smaller reactor constructions and the UV output is independent of temperature range of −20 to +70°C. The rapid start–stop procedure is not allowed and the emission is widely spread over the UV-C and UV-B ranges (Figure 31). The source is, however, very polychromatic. Estimated service time is 4000 h of operation.

2.5.3.4 Xenon Flash-Output Lamps

Mercury vapor-based emission lamps age either by repeated on–off lighting (low-pressure) or by a delay between off-to-on lighting. Direct use of electron discharge into a Penning gas (xenon at present is preferred) can make the on–off procedure more supple, however, at additional expense. (Remember that the ionization potential of xenon gas is 12.1 eV, and the energy level of the lowest excited state is 8.32 eV.)

In a proposed technology (Inovatech, Inc.), the UV emission intensity, *expressed in relative units*, is illustrated in Figure 32.

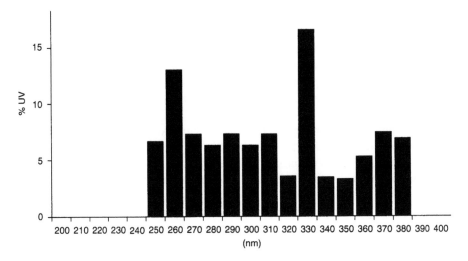

FIGURE 31 Example of UV emission of an antimony iodide-doped xenon lamp.

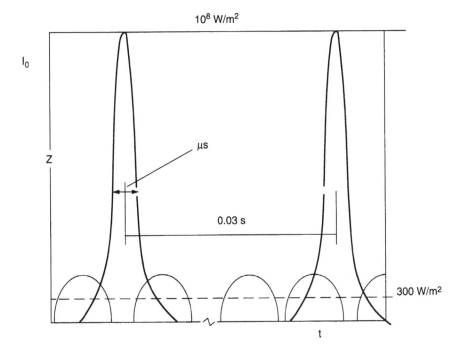

FIGURE 32 Emission of a pulsed xenon lamp.

Operated in a continuous mode of discharge, the proposed lamp emits 200 to 300 J/m^2 (depending on the wavelength; see Figure 32). Overall emission yield $W(hv)/W(e)$ is claimed to be 15 to 20%. The electron pulse last parts of a second (1 to 30 Hz; in technical versions usually 30 Hz, or 30×10^{-3} sec between pulses).

At the maximum of a single-pulse phase, claims are made that 100×10^6 J/m^2 can be obtained at the maximum pulse intensity (see Figure 32). For a single-lamp technology composed of a cylindrical lamp of 4 kW(e) (diameter not really specified in the claim, but about 2.4 cm = 1 in.), installed in a cylindrical reactor (diameter 34 cm), and operated at an estimated dose of 500 J/m^2 (over the entire emission spectrum of the xenon lamp), an estimation of operational costs is reported as 0.6 cent (U.S. \$) per 3.7 m^3. However, the costs for replacing lamps (and enclosures) must be evaluated further.

The technology certainly needs further follow-up. According to the emission in the germicidal range of UV, the potential efficiency is probable and preliminarily established (broadband UV emission). However, many unknown data remain to be developed for the technology to be fully assessed at present:

- Aging of the lamps and lamp materials (the expected lifetime of the lamps claimed to be 20×10^6 flashes or about 170 to 190 h of operation; the maximum claimed to be 1 month, but after that...?)
- Requirements for the electronics
- Emission at other wavelengths than in the range from 220 to 320 nm
- Possible secondary reactions (e.g., of nitrate ions)

As a conclusion, operation of xenon lamps in the pulsed mode is promising, but not yet fully established to be routinely applicable in current practice of large-scale continuous water treatment.

2.5.3.5 Broadband Pulsed UV Systems

In this technology, alternating current is stored in a capacitor and energy is discharged through a high-speed switch to form a pulse of intense emission of light within about 100 μsec. The ohmic heat developed ranges to a temperature of 10,000 K and the emission is similar in wavelength composition to the solar light. The expected lifetime of such lamps ranges from 1000 to 2000 h.

2.5.3.6 Excimer Lamps

A molecule (A-B in an excited molecular state, e.g., singlet excited AB^*), which is obtainable by electronic energy (not by thermal sources), on dissociation to A + B can release quantified photonic power (Figure 33).

2.5.3.6.1 Excimer Technology in the UV-C Range

Recently, a Cl_2^* excimer technology has been presented (Coogan, Triton Thalassic Technologies, Ridgefield, CT). The lamp is fed by 5 kW(e) and is claimed to emit at 260 nm with a maximum linear emission intensity of 170 W (UV)/nm (Figure 34).

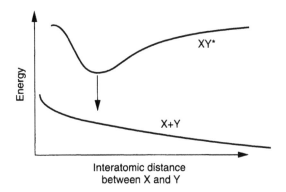

FIGURE 33 Principle of excimer emission.

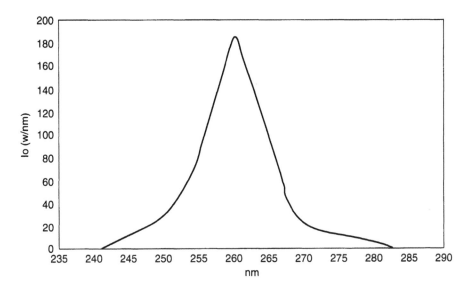

FIGURE 34 Claimed emission intensity of a 5 kW(e) Cl_2^* excimer lamp.

The total emission yield is 14% and the expected lifetime of the lamp is 6 months of operation. Gases used for the excimer technologies include also xenon–xenon chloride and krypton–krypton chloride.

Note: Excimer lamps must not be considered lasers. Excimers produce divergent radiation whereas lasers produce coherent light, at additional cost, using mirrors redirecting photons back into the gas to stimulate photon release.

2.5.3.6.2 *Excimer Technology in the Vacuum UV Range*

Most of the excimer technologies have been developed for emission in the visible range of the spectra [Braun, 1986, p. 130]. For further perspectives see Chapter 4. Also, in the case of photochemical ozone generating, the aging of the lamp and reactor materials need to be evaluated. Particularly promising are excimer technologies emitting at short wavelengths (Heraeus Noblelight Kleinostheim; also Bischof [1994]):

$$Xe_2^*, \; 172 \; nm; \; ArCl^*, \; 175 \; nm; \; ArF^*, \; 193 \; nm;$$
$$KrCl^*, \; 222 \; nm; \; XeCl^*, \; 308 \; nm$$

In general, an overview report is available from the U.S. Electric Power Research Institute (EPRI), titled "Ultraviolet Disinfection for Water and Wastewater" [1995], reporting particularly on new lamp technologies.

2.6 PRELIMINARY GUIDELINES FOR CHOICE OF LAMP TECHNOLOGY

2.6.1 Low-Pressure Mercury Lamps

- Low-pressure Hg lamps are easy to install and operate (normal take-off voltage from the mains; emission intensity can change with fluctuations in voltage of the mains).
- UV output is about monochromatic (at 254 nm) in the germicidal range (the emission at 185 nm usually is filtered by the material of the lamp).
- The relative monochromatism is not always able to initiate photochemical synergistic processes (see Chapter 4).
- No benefit is obtained by exposure of germs to other wavelengths (absorption by proteins, e.g., enzymes; see Chapter 3).
- Aging of the lamps and their materials of construction is slow. Lifetime is about 1 year.
- The lamps (plus lamp enclosures) operate at low temperatures, optimum at 40 to 42°C. At lower water temperatures, efficiency can be decreased.
- Start–stop procedure is easy. Each on–off lighting corresponds to an additional aging of 1 h.
- However, the linear emission intensity is low—ranging from 0.2 to 0.3 W (UV)/cm; consequently, the lamps are suited for germicidal action for low water flows. Treatment of high flows requires multiplication of the number of lamps installed, and as a consequence large reactor hardware.
- Doped low-pressure Hg lamps can emit at higher linear intensity (e.g., indium or yttrium doping), but at the expense of shortened lifetime (by about 50%).
- Some variants of doped low-pressure Hg lamps can emit in the range of 200 to 240 nm, which can be interesting for synergistic action (e.g., with hydrogen peroxide; see Chapter 4).
- Low-pressure lamps are emitting at nominal regime that cannot be modulated as a function of the power applied. This characteristic is not directly suited for automation of the intensity for significantly variable needs in time (e.g., variation in the flow of water to be treated).
- Low-pressure Hg lamps are readily available, with their emission spectra well established and quantified at comparatively low cost. Their more simple technology can be preferred in remote areas.
- Variants using solar energy batteries now are available.

2.6.2 Medium-Pressure Mercury Lamps

- Medium-pressure lamps that have a high linear emission intensity in the UV-C range exist (classically 10 to 15 W UV-C per centimeter; 30 W UV-C per centimeter are now possible).
- The source is polychromatic (i.e., emitting at several wavelengths), and part of the light—at least 40 to 50%—is directly useful for disinfection (see Chapter 3).
- The lamp source is operated at high temperatures (outside temperature 400 to 800°C), with lamp enclosure being of the utmost importance.
- At the same potential disinfection efficiency, medium-pressure lamps are of much smaller construction than the low-pressure Hg lamps; hence reactor design is much smaller, especially for treatment of high water flows.
- The lamps are operated at higher electrical potential (3 to 5 kV), requiring a transformer, which in turn enables modulation of the emitted intensity as a function of variable parameters of demand (e.g., water flow). This means that medium-pressure Hg lamps have much more capabilities of automation compared with low-pressure Hg lamps. (In currently available technologies of medium-pressure Hg lamps, the power input can be modulated in the range of 60 to 100% of nominal.)
- Aging of lamp material and enclosures of medium-pressure Hg lamps is faster than for the low-pressure lamps (about 4,000 vs. 10,000 h); however, improved technologies are available.
- Medium-pressure lamps can be designed as broad spectrum, emitting in a wide range of increased potential efficiency, both for disinfection and for synergistic oxidation processes.

2.6.3 Particular Lamp Technologies

To be complete, indicative data on special lamp technologies are reported in this chapter. Some of these technologies are claimed to be promising, but not yet thoroughly established in the field.

2.7 ULTRAVIOLET EMISSION YIELDS AND MODE OF CONTROL

2.7.1 Materials of Lamp Walls and Enclosures

Generally, UV lamps for water treatment are constructed of quartz, which is fragile and subject to lowered transmission by formation of deposits and slimes. Cleaning with ultrasonic techniques fails. A 1972 patent [Landry, 1972, noted in Legan, 1982] claims that certain fluorocarbons could transmit UV light. Teflon™, which is chemically inert or resistant to many products and has a refractive index close to that of water (Teflon 1.34; water 1.33 vs. air 1), has been described as having potential advantages compared with quartz [Legan, 1982]. Lamps were constructed with Teflon walls or with Teflon coatings on quartz lamps. However, the transmittance of the material, particularly on aging, has remained a subject of discussion. The data illustrated in Figure 35 [from Sagawara et al., 1984] indicate that the material is more suitable for operation in the UV-A range instead of the direct germicidal UV-C range.

2.7.2 TRANSMISSION–REFLECTION YIELDS OF OPTICAL MATERIALS

The transmission in the UV range of materials used in lamp and reactor construction is illustrated in Figure 36 (see also Figure 35).

Quartz undergoes loss in transmission by solarization, which is one of the causes of aging of the lamps; this also holds for the lamp enclosures. According to our experience (with low-pressure lamps only), in normal service quartz enclosure material can drop in transmittance at 254 nm by 50% within 20,000 to 30,000 h of service, by photochemical solarization of the material only. Scaling deposits and slime-building effects on the material need to be evaluated as well. For ozone-generating lamps, comments are reported in Chapter 4.

The following additional comments can be made [Masschelein, 1992]:

Reflection yields are indicated in Tables 4 and 5 (the tables available are most often based on the 254-nm wavelength. When the refractive index is known at other wavelengths, appropriate evaluations can be made as a function of fundamental laws (laws of Snell and Fresnel, see Section 2.7.4).

Reflectance is important in the indirect irradiation technologies based on lamps installed outside the water in devices equipped with reflectors, to direct the light either to an open channel or to a central pipe (see Chapter 3). It is particularly worth noting the high reflectance of $MgO\text{-}CaCO_3$. If such

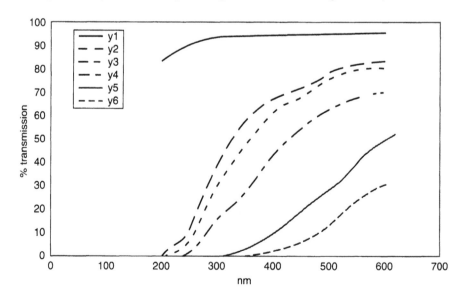

FIGURE 35 UV transmittance of Teflon vs. quartz (1: quartz 1.0 mm; 2: Teflon FEP 1000A 0.25 mm; 3: Teflon FEP 1400A 0.35 mm; 4: Teflon PFA 200LP 0.51 mm; 5: Teflon FEP 6000L 1.75 mm; 6: Teflon 900LP 2.25 mm). (From Sagawara et al., 1984.)

TABLE 4
Reflectance of UV at 254 nm

Material	Reflectance (%)
Aluminum foil	60–90
Evaporated aluminum on glass	75–90
Stainless steel	25–30
Chromium (metal)	40
Nickel (metal)	40
Oil paint (white)	3–10
Water paint (white)	10–35
Aluminum coating—paint	40–75
White plastering	40–60
Zinc oxide	5–10
Magnesium oxide	75–90
White linen	15
White cotton	30
White wool	4
Wall paper	20–30
Enamels	5–10

From technical documents of Philips Lighting, Eindhoven, the Netherlands.

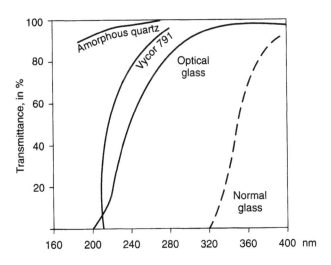

FIGURE 36 Transmission yield of optical materials.

TABLE 5
**Reflectance (in Percentage of Incident
Intensity) at 254 nm**

Material	Reflectance (%)
MgO-CaCO₃ precipitates	80–85
Magnesium oxide precipitates	70–80
Aluminum (polished)	88
Aluminum coat (mat)	75–80
Aluminum foils	73
Plaster (white hospital)	40–60
Chromium (metallic)	45
Nickel (metallic polished)	38
White paper	25
Inox (IASI-304-mat) and AISI 316	≤25
Water paints	10–30
White porcelain	5
Normal glass	4
Quiet water surface	4
Polyvinyl chloride (PVC)	1
Miscellaneous (organic) precipitates	1
Black optical paint	1

From technical literature summarized by Masschelein, 1992
to 1996.

---

precipitates occur on the lamps (*or on the enclosures*), loss in irradiation
yield can be important. This is a significant point for maintenance.

2.7.3 PRECIPITATION OF DEPOSITS (SLIMES)

The precipitation of deposits on direct contact of lamps with water is illustrated in
Figure 37.

The total amount of precipitated mineral salts on direct exposure of the lamp to
the water can range between 2.5 and 15 meq/m^2 of external lamp surface [personal
observations]. Predominant in the composition of the deposits are usually Ca or Mg
(between 30 and 80%), but in flocculated water also 20 to 30% of total as Fe or Al
precipitates and miscellaneous debris. In groundwaters containing Fe and Mn, these
minerals can represent 10 to 40% of the total. All this depends on the overall composition of the water.

However, general conclusions are:

- A lamp enclosure is necessary for continuous operation.
- This enclosure should allow thermal dissipation at the "true" lamp wall.
- Longitudinal flow of the water along the lamps, which exposes the water
  to precipitating effects, can be more critical than the transverse mode of
  installation.
- Procedures for cleaning, continuous or intermittent, are necessary.

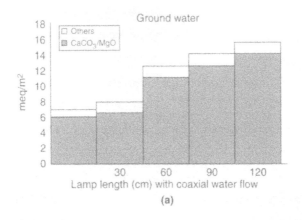

(a)

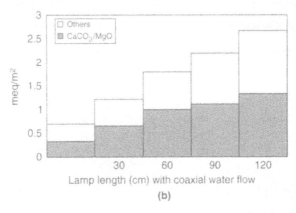

(b)

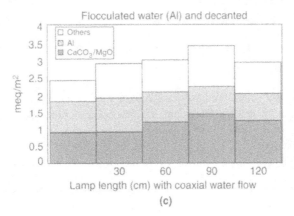

(c)

FIGURE 37 Precipitation of mineral deposits from the water on UV lamps (a) groundwater; (b) groundwater containing iron and manganese; (c) surface water, coagulated, flocculated, and settled with alum.

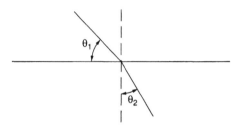

FIGURE 38 Transmission reflectance of water according to Snell's law.

Conclusion—It is necessary to place the lamps in a quartz enclosure tube to compensate for the drop in emission yield of low-pressure Hg lamps at low water temperature and to prevent mineral precipitations on the lamp walls of medium-pressure Hg lamps.

2.7.4 TRANSMISSION–REFLECTANCE BY WATER

According to Snell's law, the refractive properties of a surface are related as follows:

$$n_1 \sin \theta_1 = n_2 \sin \theta_2$$

where n_1 and n_2 are the refractive indices of the two media, and the angle of refraction θ_2 is smaller than the angle of incidence θ_1 (Figure 38).

The law of Fresnel relates the refractive indices to reflectance and transmittance properties of materials. The basic relation is $T = (1 - R)$ (T = transmittance; R = reflectance). It is interesting to note that the reflectance for UV-C at the air–quartz interface remains on the order of 4 to 5% of emitted light as long as the angle of incidence (θ_1) remains lower than 50° (Figure 39).

Similarly, the reflectance of a plain water surface (at 254 nm) is about 4%. As a consequence, one must consider a 4 to 5% loss of intensity by reflectance in the transmission of the UV light, either by a quartz enclosure of the emitting lamp with the enclosure space containing air or by direct irradiation of an open water surface as in channels, for example.

In practice, the transmittance of light by a 1-cm layer thickness is often considered; this property is easily measured with a standard spectrophotometer. These aspects are very important for reactor design and are described in more detail in Chapter 3.

2.7.5 RADIOMETRY

Precise radiometric determinations are to be made by *specialized laboratories in the field*, instead of water utilities. The most classic method of measuring the radiant intensity of a source is to place it in the center of an integrating sphere. A detector located at the surface of the sphere will indicate the power of the radiation received on the sphere, and therefore the total radiant intensity of the source. The results usually are expressed as tabulated values in catalogs of *irradiance* per unit surface at 1-m distance from the source (W/m^2), emitted per steradian, that is, in a solid angle corresponding to 1 m^2 located on a sphere of 1-m radius, W = $4\pi R^2$ (for $R = 1 - $ m).

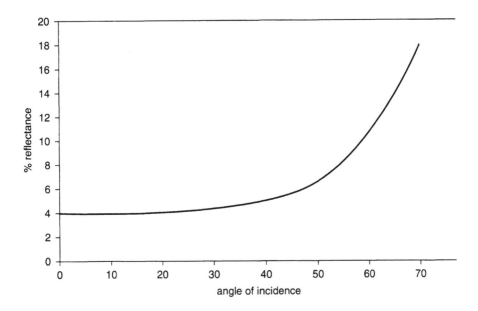

FIGURE 39 Reflectance (R) of UV as a function of angle of air–quartz incidence.

Under practical conditions of measurement, still in an ideal integrating sphere, the reflectance of the material of the sphere needs to be considered, and one has $W = 4\pi R^2/(r/(1 - r))$. The construction of such spheres that can measure the output of long cylindrical lamps has been, and still is, a challenge for standardization institutes.

Important for practice is that the values for intensity (irradiance) measured by this method and tabulated accordingly, generally concern the overall radiant intensity emitted in a given domain of UV (e.g., 180 to 300, 180 to 400, or 220 to 400 nm). The values do not specify the spectral distribution of the energy. For this purpose, one needs a spectral radiometer, a device for selection of wavelengths that includes optical filters, separational monochromaters (such as prisms or diffraction gratings), and a more or less specific detector like a photoelectric cell or a photodiode and thermoionic detector. An accurate standard source also is necessary.

It is important, however, that the utilities are informed about a certain number of essential principles and are able occasionally to carry out simplified tests for control and maintenance of the equipment. Several instrumental arrangements are available on the market. With these instruments it is possible to obtain emission intensities of light sources as indicated hereafter.

2.7.6 OPTICAL FILTERS

Photosensitive papers can be used to some extent for absolute and to a large extent for relative measurements (one source vs. another) of the radiant flux emitted by a source. If a spectral domain of particular interest for a given application emitted by a source (e.g., 220 to 280 nm) is delimited by an appropriate light-filtering system, the blackening of a photographic film can be used as a measure of potential efficiency of the application. Between threshold and saturation level of exposure, the optical density of the exposed film is correlated linearly to the log of the exposure dose. By substituting a thin layer of water contaminated with viable organisms to the same irradiation, a direct dose-to-effect measurement can be established.

As for the photon-selecting (filtering) system, many solutions are possible. In research on water treatment and also for on-site testing, transmittance filters are most appropriate. Some examples are illustrated in Figure 40.

Other technologies such as interference filters and dielectric filters are used more in specialized laboratories (see Murov [1973]).

2.7.7 Spectral Radiometry (Photocells)

This technique can be applied on-site if calibrated photocells are available for the wavelength or the wavelength zone under consideration. Most of the radiometers used in current practice are photonic cells. Such cells involve a UV-sensitive cathode that converts the incident light intensity into electrical current (photoelectric effect). Such actually available detectors are very sensitive, however, often not specific for a given wavelength.

2.7.7.1 Specific Photocells

Such cells are available for the 254-nm band. Calibrated photocells are available in two types, cylindrical and cosinusoidal (Figure 41).

Simplified photocells of the cylindrical type are suitable for continuous semi-quantitative monitoring of continuous operating equipment in water treatment (see Chapter 3). For the determination of the nominal emission intensity, the use of calibrated cosinusoidal-type photocells is preferable. In the experimental setup indicated, the measured power is given by:

$$P = 2\int_{90°}^{0}\int_{0}^{2\pi} I(\theta)d\theta\, dA(\theta)$$

$$\int_{0}^{2\pi} I(\theta)d\theta = 2\pi a(\theta)I(\theta)$$

$$dA(\theta) = rd\theta$$

Hence:

$$P = 2\int_{90°}^{0°} 2\pi a(\theta)I(\theta)rd\theta$$

and:

$$\frac{1}{2}\frac{dp}{d\theta} = 4.022\, I\sin\theta \quad (\text{for } r = 0.8m)$$

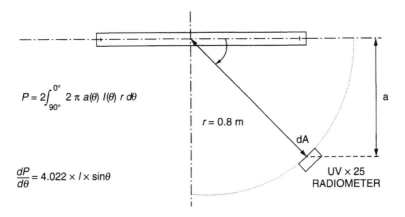

$$P = 2\int_{90°}^{0°} 2\pi\, a(\theta)\, I(\theta)\, r\, d\theta$$

a

$r = 0.8\ m$

dA

$$\frac{dP}{d\theta} = 4.022 \times I \times \sin\theta$$

UV × 25
RADIOMETER

This technique is also adequate for relative measurement of the power loss of a given lamp through aging. Graphic integration of the area under an experimental curve is obtained by plotting the cumulative $dP/d\theta$ value as a function of θ between 0° and 90°, by rotating the photocell in a sector of 90° in a plane along the axis of the tubular lamp, and by recording the power per radiant; by these means one can construct graphs as in the example given in Figure 42. The area below the curve corresponds to half of the nominal power emitted by the lamp in the photocell detection zone. If the lamp is not cylindrical (e.g., U shaped), an additional integration is required by rotating the detector by zero to 360° around the main axis. The formula then becomes:

$$P = 2\int_{90}^{0}\int_{0}^{360} 2\pi a(\theta)I(\theta)r^2 \sin(\theta)$$

This method and its required simple hardware make it very useful for occasionally controlling the aging of the lamps and for checking the transmission yield of the (quartz) enclosures.

2.7.7.2 Nonspecific Photocells

Most of the photocells have a nonspecific response vs. wavelength. A typical example is illustrated in Figure 43. Some phototubes can have more narrow detection limits, but generally one has to rely on cut-off filters. Broadband phototubes often are used for continuous survey of reactors as installed (see Chapter 3).

2.7.8 ACTINOMETRY

Fundamentally, the primary quantum yield has a maximum value of unity (Stark–Einstein law). This means that when one photon is absorbed by a molecule, it induces a single change in the molecule. However, the primary photochemical reaction can be followed by further reactions, photochemical or not. The overall quantum yield is defined by the number of molecules transformed per number of photons absorbed (e.g., in moles per Einstein).

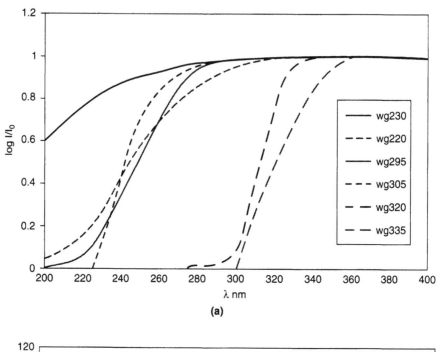

(a)

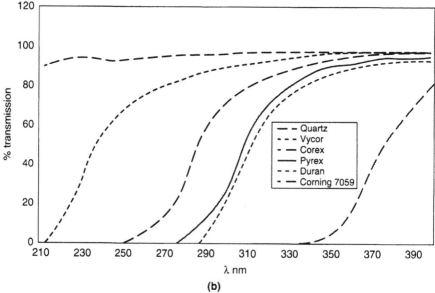

(b)

FIGURE 40 (a) and (b) Transmittance of optical glass filters. (From documents of Scott, Mainz, Germany.)

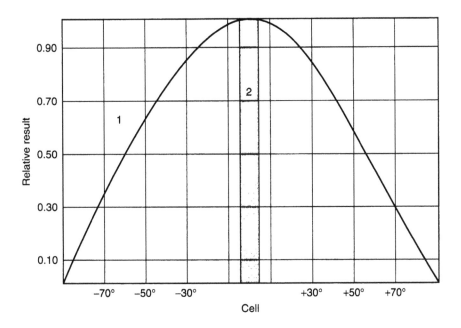

FIGURE 41 Detection profile of cylindrical (2) and cosinusoidal (1) photocells. (From Masschelein, 1986; 1992; 1996.)

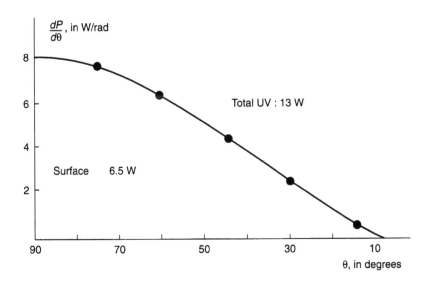

FIGURE 42 Graphic integration of nominal power measured by photocells. (From Masschelein, 1992.)

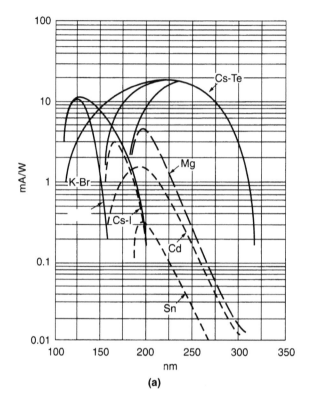

(a)

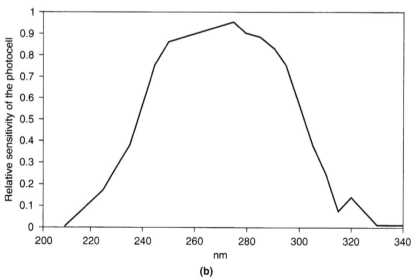

(b)

FIGURE 43 (a) and (b) Typical response curve of a commercial photocell in UV range. (From documents of Hamamatsu, Japan.)

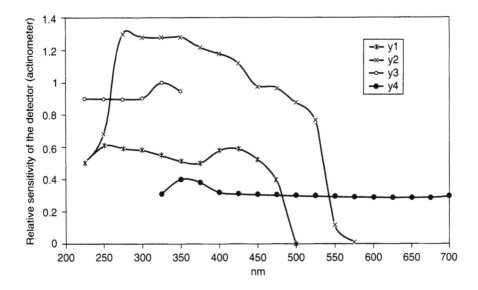

FIGURE 44 Quantum yields of actinometers. (From general literature and also from documents of Philips Lighting, Eindhoven, the Netherlands.)

The actinometry supposes a quantitative and reproducible relation between the number of photons of a given frequency irradiating a reagent and a known single photon-induced chemical transformation. In experimental determinations it is preferable that all photons are absorbed.

2.7.8.1 Mineral Salt Actinometers

Three classical actinometers are often used for the UV region: uranyl salts coupled with oxalic acid, potassium ferrioxalate, and malachite green leucocyanide (Figure 44).

U(IV) is oxidized by UV photons to U(VI), which in turn oxidize oxalic acid. Determination of the oxalic acid before and after irradiation is correlated with the number of photons absorbed by the system, with the quantum yield 0.6 for the wavelengths active in disinfection:

$$I \times 4.7 \times 10^5 = P \quad \text{(in watts of UV if monochromatic at 254 nm)}$$

Potassium ferri(III) oxalate is transformed into ferro(II) oxalate and the concentration is correlated with the number of photons absorbed. In the region of interest in water treatment, the quantum yield to be considered is 1.23 (see Figure 44). The lowering of the quantum yield at wavelengths under 250 nm makes the ferrioxalate actinometer less adaptable for the evaluation of germicidal efficiency of a lamp compared with the uranyl-oxalic acid method.

Irradiation of malachite green with UV gives a green color with a measurable absorption at 622 nm. The quantum yield is about unity in the range 200 to 300 nm.

One advantage of the actinometric methods is that apparently they are less hardware-dependent than are the radiometric methods. They can be operated in a laboratory with standard equipment available and even applied to reactors on real scale. Drawbacks are:

- Lack of specificity vs. wavelength
- Possible inner effects, (i.e., the secondary absorption of photons by the reaction products of a first reaction)
- Very strict requirements for observing the experimental conditions concerning pH, concentration ranges, etc.

Recommended are the experimental procedures described originally and, when necessary, adapted to the reactor volumes applied:

- For the uranyl oxalate method [Leighton and Forbes, 1930]
- For the potassium ferrioxalate method the original procedures described by Hatchard and Parker [1956]
- For the malachite green leucocyanide method by Harris and Kaminsky [1935]; potential effects of light in the visible range that must be considered; therefore, the operations to be conducted in the "visible dark"

2.7.8.2 Iodide-Iodate UV Actinometer

In the iodide-iodate actinometer [Rahn, 1997] the photolysis of iodide ion in the presence of iodate ion, which acts as an electron acceptor, corresponds to following overall reaction:

$$8I^- + IO_3^- + 3H_2O + (x)\ hv = 3(I_3)^- + 6OH^- + 6H^+$$

The reaction is considered by hypothesis to proceed through different intermediates (I°; e-aq.; OH^-; H_2O_2, etc.); however, the overall stoichiometric result is as indicated earlier.

The concentrations utilized are 0.6 mol/l KI and 0.1 mol/l KIO_3. The solution absorbs in the region of 200 to 300 nm and the reaction is not directly influenced by visible light. The concentration of the triiodide anion can be measured at 352 nm ($A = 26,400$ l/mol·cm). The quantum yield distributed over the entire range is 0.75. The response depends on the concentration of the reactants.

Recommended preparation of the actinometer solution is as follows: in 10 mL ultrapure water (e.g., ISO Nr 3686, 1987) dissolve in sequential order: 1 g KI, 0.214 g KIO_3, and 0.038 g borax (to make the solution 0.01 molar in $Na_2B_4O_7$). Fresh preparation before use is required. The actinometer solution can be controlled as follows: pH about 9.25; A (l/mol·cm) 0.6 ± 0.03 at 300 nm and 0.04 at 352 nm. The absorbance of the actinometer without UV exposure is negligible at wavelengths higher than 330 nm. However, the absorbance increases with temperature, about linearly (at least between +10°C and +45°C) by a factor of +14% per increment of 10°C; hence, a correction for temperature may be required.

The absorbance (base$_{10}$) at 20°C of the triiodide ion in water is summarized in Table 6.

TABLE 6
Absorbance (Base$_{10}$)
at 20°C of the Triiodide
Ion in Water

λ (nm)	A (l/mol·cm)
330	15,500
340	29,500
345	23,000
352	24,600
375	16,751
400	6,196
425	2,773
450	1,388

2.7.8.3 Persulfate *tert*-Butanol UV Actinometer

Persulfate ion in aqueous solution is decomposed by UV light of wavelengths lower than 300 nm [Mark et al., 1990]. Visible light does not interfere. The basic reactions are:

$$S_2O_8^{2-} + h\nu \rightarrow 2SO_4^{-\bullet}$$

$$SO_4^{-\bullet} + (CH_3)_3COH \rightarrow H^+ + SO_4^{2-} + \bullet CH_2C(CH_3)_2OH$$

In the absence of dissolved oxygen, the *tertiary*-butyl radicals dimerize, and the dimers are not able to act as supporters of chain reactions. In the absence of dissolved oxygen, the quantum yield Φ is 1.4. In the presence of dissolved oxygen, secondary reactions can occur:

$$\bullet CH_2C(CH_3)_2OH + O_2 \rightarrow \bullet O_2CH_2C(CH_3)_2OH$$

These pseudoradicals are not directly chain carriers in the reduction of peroxydisulfate, but they can further build up oxygen radical ions [Buck et al., 1954], which are able to promote further reductions of peroxydisulfate ions according to: ($O_2^{-\bullet}$ + $S_2O_8^{2-} \rightarrow O_2 + SO_4^{2-} + SO_4^{-\bullet}$. The sulfate anion radical is able to further oxidize *tert*-butanol radicals. In such a case, the quantum yield can be in the range of 1.8 [Becker, 1983]. *Consequently, control of dissolved oxygen is important to apply this method.*
Recommended concentrations are [Winter, 1993]:

- Potassium peroxydisulfate: 0.01 mol/L (2.7 g)
- *Tertiary* butanol: 0.1 mol/L (7.4 g)
- Dissolved in 1 L distilled water
- Saturated with dissolved oxygen vs. air by bubbling for 30 min

The use of freshly (daily) prepared actinometer solution is recommended.

2.7.8.4 Uridine Actinometry

Uridine has an absorption spectrum that partly matches well with the general absorption spectrum of DNA and covers the range of 200 to 300 nm (Figure 45).

In aqueous solution, the sodium salt of uridine has a maximum absorption at 267 nm, which is decreased by UV irradiation. Decrease in absorbance is not linear as a function of the dose and remains very low at conventional doses for disinfection. However, the technique can be useful for evaluation of high-irradiation doses, such as for direct photochemical processes. Photohydration is advanced as an interpretation.

Reported data are in following order of magnitude, in terms of decrease of absorbance at 267 nm [Linden, 1999]:

$$\text{In terms of J/m}^2; \ 3 \times 10^2, \text{ about nil; } 3 \times 10^3, \ -8\%;$$
$$3 \times 10^4, \ -13\%; \ 3 \times 10^5, \ -45\%; \ 3 \times 10^6, \ -97\%$$

Therefore, the method is promising, but much further development is required, particularly on calibrating methods for quantifying high-irradiation doses. See also [Linden and Darby, 1997].

2.7.8.5 Hydrogen Peroxide Decomposition as an Actinometric Check-Control Method

The most commonly accepted initial reaction in the photodecomposition of hydrogen peroxide in water is:

$$H_2O_2 + h\nu = 2OH\bullet$$

The reaction is first order and the kinetic decay constant can be correlated as:

$$k = (2.3 \times A \times \Phi \times L \times r \times I_0)/V$$

The meaning of the symbols is as usual, with r the reflectance of the (cylindrical) reactor wall [Guittonneau et al., 1990].

In the concentration range of some milligrams per liter, the quantum yield Φ (at 20°C) is reported as 0.97 to 1.05 [Baxendale and Wilson, 1957]; however, the quantum yield depends on temperature [Schumb and Satterfield, 1955]. The determination of the photolysis ratio of hydrogen peroxide and the ease of measurement of its residual concentration after exposure makes it an appropriate method for control of the constancy of operational experimental conditions by a rapid "morning-check" preceding a series of experiments.

More fundamental aspects of UV photolysis of hydrogen peroxide are commented on further in Chapter 4.

2.8 ZONAL DISTRIBUTION OF EMITTED LIGHT

Lamps have a zonal distribution of emitted light intensity. Some indications can be found in Figures 46 and 47. Lamps present a drop in intensity at the electrode ends (Figure 46). Because of the configuration of the electrodes and their location in space, cylindrical lamps often have an uneven cylindrical distribution of emitted intensity.

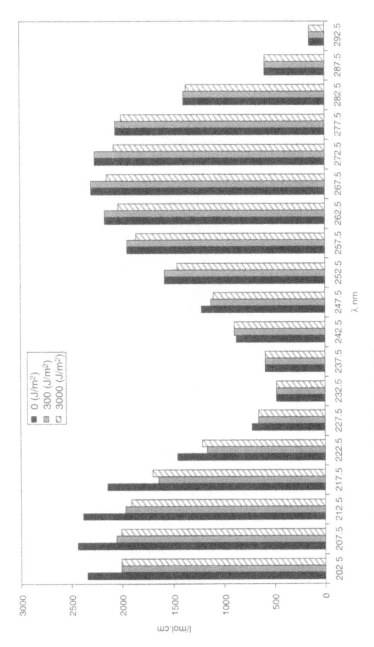

FIGURE 45 Absorption spectrum of uridine in the UV range.

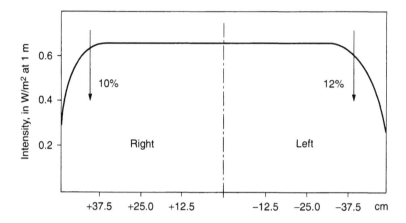

FIGURE 46 Drop in intensity at the electrode ends. (From Masschelein, 1996b).

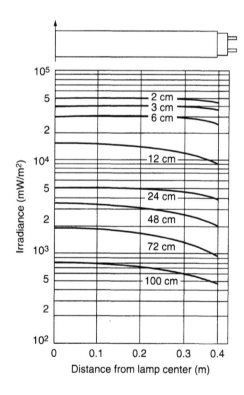

FIGURE 47 Drop in UV intensity at the lamp ends as a function of the distance below the lamp. (From documents of Philips Lighting, Eindhoven, the Netherlands.)

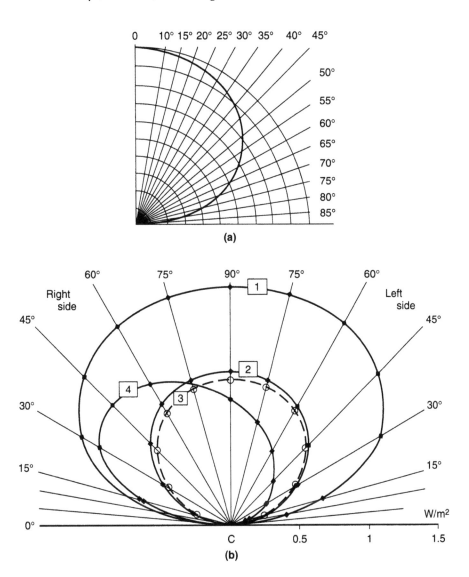

FIGURE 48 Illustration of emission distribution (schematic polar diagram).

More classical tubular lamps have a zonal distribution. Figure 48 shows a typical example.

The drop in intensity at the lamp ends also depends on the distance below the lamp, as illustrated in Figure 47.

Because mercury vapor UV emitters are plasma emitters and not really true point sources, the intensity (emitted by a small section) varies with the apex angle to normal. This aspect is important in reactor design and geometry of location of the lamps in multilamp reactors (see Chapter 3).

For longer and perfectly cylindrical lamps, the radial model (i.e., in which the intensity is emitted identically and orthogonally in all directions from the lamp surface) often ends in good approximations, although it is obviously erroneous from the fundamental point of view [Phillips, 1983, p. 368]. Flat-shaped lamps (see Figure 25) are claimed to emit about two-thirds of the total intensity on the flat side of the lamp wall. This is significant for the design of the reactors using this lamp technology. U-shaped lamps as illustrated in Figure 10 and bend-type lamps as illustrated in Figure 17 integrate a compromise solution between cylindrical and flat-shaped lamp technologies exploiting the zonal distribution of emission, also called polar distribution. Information on all these very fundamental aspects should be made available for the user.

3 Use of Ultraviolet Light for Disinfection of Drinking Water

3.1 INTRODUCTION

The number of drinking water systems relying on ultraviolet (UV) irradiation for disinfection of the water, at present, is estimated to be about 3000 to 5000. The use of the technique is probably much higher in number, because these applications are often not completely recorded:

- Point-of-use of the system on household scale, camp grounds
- Recreational and body health applications
- Applications in risk zones such as hospitals, nurseries, and schools in remote areas
- Use in food processing industries such as breweries and soft drinks industries
- Use on boats, ships, and railway trains

Bactericidal effects of radiant energy from sunlight were first reported in 1877 [Downes and Blunt, 1877]. However, thanks to the absorption by atmospheric ozone, the part of UV from sunlight that reaches the surface of the earth is merely confined to wavelengths higher than 290 nm. The technical use of UV made progress by the discovery of the mercury vapor lamp by Hewitt [1901] and the drinking water of the city of Marseille in France was disinfected with UV light as early as 1910.

The reliable operation and functioning of 5000 plants cannot be ignored in spite of some suspicions or objections that have been formulated (to be commented on in this chapter). Among them is the absence of active residual concentration in the treated water [Bott, 1983]. This point has pros and cons, but because no on-site storage of chemicals is required, the risk for the operators is eliminated and the safety measures and equipment for handling chemicals are not needed. In remote areas, transportation problems may be solved as well. Versions operated on the basis of solar photoelectric generators are developed now and are available.

Since late 1979 in the area of Berlin, Germany, the treated water has not been postchlorinated.

The question of maintaining an active residual in the water in the distribution system certainly remains a subject of option, debate and, local circumstances (i.e., overall water quality). Although not a central point of present information, this matter should not be ignored.

3.2 GERMICIDAL ACTION

3.2.1 GERMICIDAL ACTION CURVES

According to the Grothius–Draper law, only absorbed photons are active. Considering disinfection with UV light fundamentally to be a photochemical process, the UV photons must be absorbed to be active. This absorption by cellular material results from absorption by proteins and by nucleic acids (DNA and RNA). The respective absorbances are indicated in Figure 49.

The overall potential disinfection efficiency of UV-C is illustrated in Figure 50.

3.2.2 MECHANISM OF DISINFECTION

The germicidal efficiency curve closely matches the UV absorbance curve of major pyrimidine components of nucleic acids, as illustrated in Figure 51.

The absorption in the UV-C range of nucleic acids roughly corresponds to the UV absorption by the pyrimidine bases constituting part of the nucleic acids. From photochemical irradiation of the different pyrimidine bases of nucleic acids, the isolated products are principally dimers, mainly from thymine and secondarily from cytosine. The relative germicidal action curve as a function of the absorbance is reported in Figure 52.

Bacterial decay is considered to occur by lack of capability of further multiplication of organisms, for example, with damaged nucleic acids. Possible repair mechanisms have been taken into consideration as well. Various mechanisms of repair of damaged nucleic acids can occur (Figure 53 [Jagger, 1967]).

The thymine dimer absorbs light (e.g., in the visible range [blue light]), a characteristic that is supposed to restore the original structure of the damaged nucleic acids. (The question remains open as to whether modified DNA cannot induce [plasmids] modified multiplications if the general protein structure of the cell is not destroyed as well; see Figures 50(b) and 54.)

Enzymatic repair mechanisms are described involving a UV-exonulease enzyme and a nucleic acid polymerase: [Kiefer, 1977; Gelzhäuser, 1985]. The process supposes an excision of the dimer followed by a shift in one of the wraps of the nucleic acid.

The repair of bacteria after exposure to UV-light is not universal. Some organisms seem not to have the capability of repair (*Haemophilus influenzae, Diplococcus pneumoniae, Bacillus subtilis, Micrococcus radiodurans,* viruses); others have shown the capability of photorepair (*Streptomyces* spp., *E. coli* and related enterobacteria, *Saccharomyces* spp., *Aerobacter* spp., *Erwinia* spp., *Proteus* spp.) [U.S. EPA, 1986]. Similar data have been reported (Bernhardt, 1986). The conclusion of the latter contribution was that to avoid photorepair, an additional dose was required vs. the strict Bunsen–Roscoe law concept. Viruses as such, when damaged by UV irradiation, have no repair mechanisms.

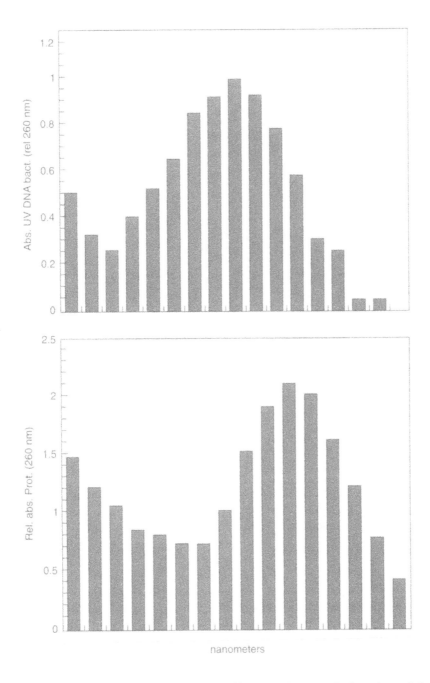

FIGURE 49 UV absorbance of cellular matter of bacteria (histograms by 5-nm intervals from 215 to 290 nm).

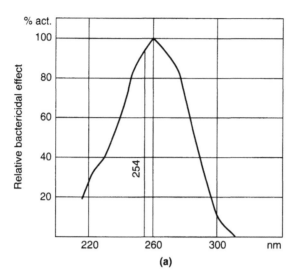

(a)

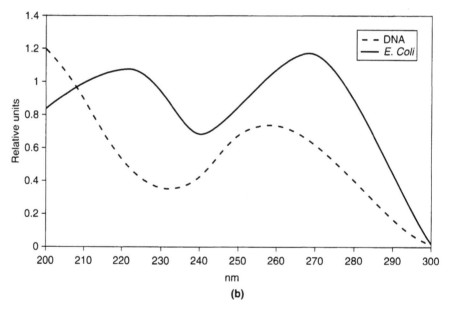

(b)

FIGURE 50 (a) Germicidal efficiency distribution curve of UV based on maximum at 260 nm; (b) overall absorbance of *Escherichia coli* vs. DNA.

After exposure to higher doses, coliform bacteria exhibit less or no repair at all [Lindenauer and Darby, 1994]. Also, for photorepair, exposure to light (300 to 500 nm) must occur a short time after exposure to germicidal light (within 2 to 3 h) [Groocock, 1984]. More complete photorepair may last up to 1 week for *E. coli* [Mechsner and Fleischmann, 1992].

Further information on more frequently observed repairs in treated wastewaters is given in Chapter 5. However, the investigations on repair after UV action generally

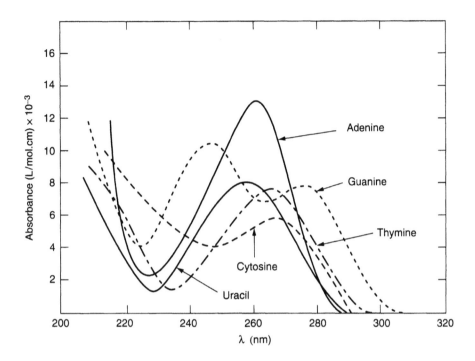

FIGURE 51 UV-C absorptivity of pyramidine bases. (According to data reported by Jagger, 1967.)

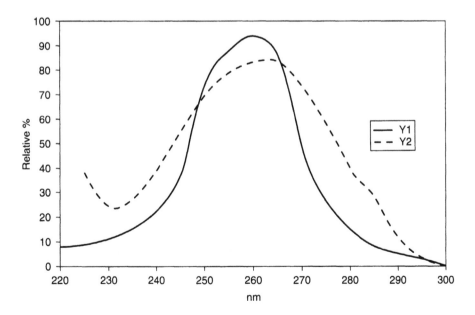

FIGURE 52 Possible relation between germicidal efficiency and absorption of UV light by the thymine component of nucleic acids.

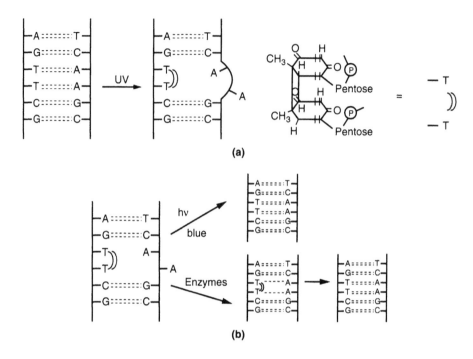

FIGURE 53 (a) Schematic of dimerization of the thymine base and possible repair mechanisms. (b) Possible repair mechanisms of UV-injured nucleic acids.

were made after exposure to low-pressure monochromatic UV lamps. After exposure to broadband UV lamps, which are able to induce more general cellular injuries, no conclusive evidence of repair has been produced as yet. This point may still need further investigation.

As a preliminary conclusion, the enzymatic repair mechanism requires at least two enzyme systems: an exonuclease system as, for example, to disrupt the thymine–thymine linkage, and a polymerase system to reinsert the thymine bases on the adenosine sites of the complementary strain of the DNA. However, on appropriate irradiation, the enzymes seem to be altered as well.

Aftergrowth has not been observed in waters distributed through mains (i.e., in the dark) as long as the dissolved organic carbon (DOC) remains low (e.g., lower than 1 mg/L) [Bernhardt et al., 1992]. However, further investigation is under way.

In addition, the literature approach often neglects the possible effects of polychromatic UV-C light on proteins, inclusive of enzymes as potentially involved in repair mechanisms.

3.2.3 Potential Effects on Proteins and Amino Acids

Proteins absorb UV-C light as illustrated in Figure 49, principally by the amino acids containing an aromatic nucleus (i.e., tyrosine, tryptophan, phenylaniline, and cystine-cysteine). Peptides containing a tryptophan base have been shown to undergo photochemical changes with conventional UV irradiation by low-pressure mercury lamps

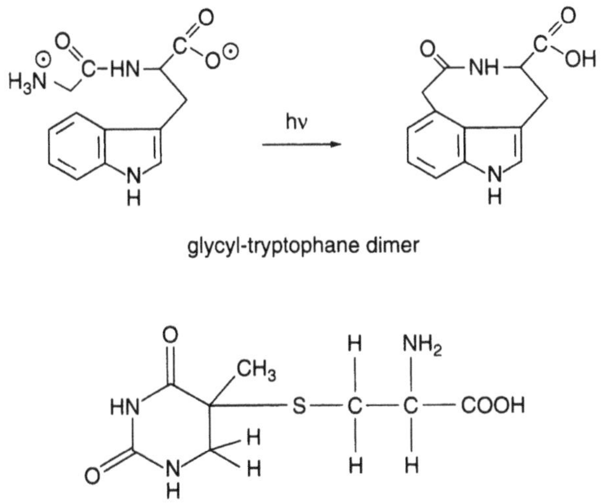

FIGURE 54 Example of photochemical reaction of proteinaceous matter.

[Aklag et al., 1990]. Among them the glycyl-tryptophan dimer (unit of proteins) has been shown to produce a condensed molecule. No mutagenic activity (Ames test), is associated with this structural modification. Other reactions are DNA protein cross-links as, for example, in Figure 54 with cysteine (according to Harm [1980]).

Thus far, the investigations have often been concentrated on low-pressure Hg lamp technologies emitting essentially at the 254-nm wavelength. By considering the emission spectra of medium-(high-)pressure lamps (see Chapter 2), the importance of photochemical changes in proteins may become of higher priority (e.g., in deteriorating capsid proteins of viruses and constitutional proteins of parasites). Reactions on such sites are indeed considered to be important in disinfection with chemical agents such as chlorine and chlorine dioxide. The question is actively under investigation, particularly in the field of inactivating organisms other than bacteria.

3.2.3.1 What Can Represent UV Absorbance of Bacterial Proteins?

By using enterobacteria as an example, the dry body mass ranges 10^{-12} to 10^{-13} g, about half of which is carbon mainly in proteins and protein-related lipids. By taking as an average 5×10^{-13} g per bacterium and considering an arbitrary concentration of 6.02×10^{6} bacteria per liter (or 10^{-17} mole-bacteria per liter), 3×10^{-6} to 6×10^{-6} g/L of cellular proteins results (in terms of mass of carbon). The molar mass of cellular proteins ranges from 10,000 to 50,000 (exceptionally up to 100,000), which equals 10 to 100 kD. By taking $25,000 \pm 15,000$ as an assumption, by considering that the absorbance of cellular proteins is in the range of about 100 L/mol·cm at 254 nm, and by roughly assuming that most of the carbon is linked to cellular proteins, this results in a potential optical density at 254 nm (of the bacterial population as given before) of about $2.4 \pm 1.5 \times 10^{-8}$ cm^{-1}. However, the overall absorbance of cellular proteins increases at shorter wavelengths (≤220 nm) to attain 4000 to 5000 l/mol·cm, which is about equal to the absorbance of single-stranded DNA (see Figure 49).

Also, some individual amino acids absorb strongly in the UV range. For example, tyrosine presents a maximum at 220 nm (8200 L/mol·cm) and a secondary maximum at 275 nm (1450 L/mol·cm); and tryptophan, at 220 nm (33000 L/mol·cm) and at 275 nm (5600 L/mol·cm). Other vital components like cytochrome c in its oxidized form absorb strongly in the UV-C range.

3.2.3.2 What Can Represent Cellular DNA (RNA) Concentration in Terms of Quantitative Absorption of UV?

The size of DNA usually is reported in terms of thousands of kilobases (kb), which represent the length of 1000 units of base pairs in a double-stranded nucleic acid molecule (for bacteria), or 1000 bases in a single-stranded molecule (bacteriophages, viruses). Typical values are viruses, 5 to 200 kb; phages, 160 to 170 kb; E. coli, 4,000 kb (general bacterial mycoplasma, 760 kb); yeasts, 13,500 kb; and human cells (average), 2.9×10^6 kb.

When considering E. coli and the intranuclear part of DNA, 4000 kb represent about 2.6×10^6 kDa (1 kb = ±660 kDa and 1 Da = 1.68×10^{-24} g); this means ±4.4 × 10^{-15} g DNA per bacterium. In the example of a population of 6×10^6 bacteria per liter, the concentration represents about 2.6×10^{-8} g intranuclear DNA per liter. At an average molar mass per base pair of 820, the example ends at about 3×10^{-11} mole base pairs per liter, or 1.2×10^{-7} moles intranuclear DNA per liter of water.

The absorbance of DNA *isolated* from E. coli in the UV-C range is illustrated in Figure 49. Isolated single-strand DNA presents a maximum at 260 nm of about 5200 l/mol·cm; and isolated double-helical DNA, 3710 L/mol·cm. (Some inner-shielding effect occurs in the double-stranded DNA.)

Note: All these values reported are for isolated DNA and not cellular DNA. Taking 4500 L/mol·cm as a preliminary value, for a concentration of 1.2×10^{-7} mol/L, this results in an estimated optical density (at 254 nm) of 5.4×10^{-3} cm^{-1}.

3.2.3.3 Conclusions

- DNA and its constitutive bases (see Figure 51) have strong absorbances around 254 nm, but overall in the range of 200 to 300 nm. Cellular proteins, more abundant in the living cell structure, absorb more at lower wavelengths.
- Measurements of absorbances are based on isolated material and not within the real cell structure in which the intranuclear DNA is protected by the general matter of the cells.
- The absorbance of both proteins and DNA is weak, essentially transparent to UV.
- As such, the *exposure dose* translates into the *probability* of a determinant *deactivating or killing hit* of vital centers of a cell.
- However cellular proteins, although generally less absorbent, may be a critical step to overcome, as for example, alteration of the capsid enzymes

necessary for the penetration of viruses or parasites into host cells. The surprising efficiency of medium-pressure broadband multiwave UV in deactivating parasites may be found in such photochemical reactions.
- Viruses and parasites rely on proteolytic enzymes to penetrate the host cells.
- The potential efficiency of polychromatic lamps (emitting in the range of 200 to 300 nm) vs. the more classical monochromatic lamps (essentially emitting at 254 nm) must be taken into consideration in the evaluation of the overall efficiency. More permanent disinfection can be achieved in the field with medium-pressure multiwave lamps.

Further comments—As described in Section 1.1, the direct disinfecting effect of sunlight is not strong enough to achieve direct disinfection of water. However, the *total* intensity of the solar irradiation at the surface of the earth is evaluated as 320 W/m^2 (average). In more specific regions, UV A/B medium-pressure Hg lamps can emit locally much higher intensities than the general solar irradiance (see Figure 22). In 1952, it was discovered that quanta above 300 nm up to the visible light region could inhibit the capability of multiplication of microorganisms [Bruce, 1958]. The killing effect has been considered to result from the formation of singlet excited oxygen in the cytoplasm [Torota, 1995]. As a conclusion, photons of wavelengths higher than 300 nm can contribute sigificantly to the decay of microorganisms by the absorption of chromophores other than nucleic acids. Leakage of cellular ions resulting from cell damage has been advanced as an explanation [Bruce, 1958]. The question is analyzed and commented on by Kalisvaart [2000].

3.2.4 EVALUATION OF GERMICIDAL EFFICIENCY OF LAMPS

At 254 nm, which is the main wavelength emitted by the low-pressure mercury lamp, the potential efficiency is in the range of 95% (see curve in Figure 50). Because low-pressure mercury lamps emit about 80 to 85% at that wavelength, the potential efficiency is 75 to 80% of the total emitted UV-C radiation.

Medium-(high-)pressure mercury lamps and similar technologies (Sb lamps) emitting a polychromatic spectrum must be evaluated by matching the emission spectrum to the germicidal action curve. Therefore, Meulemans [1986] has developed a histogram method, on the basis of integrating the potentially effective germicidal power in the 210 to 315-nm range by steps of 5 nm.

$$I = \text{Total potentially germicidal emitted power in the 210 to 315-nm range (watt)}$$

$$I(\lambda) = \text{Power emitted in a 5-nm segment (watt)}$$

$$S(\lambda) = \text{Potential efficiency coefficient in each 5-nm segment of the germicidal curve}$$

$$\Delta\lambda = \text{5-nm segment interval of integration}$$

$$I\,(\text{watt}) \,=\, \Sigma[I(\lambda) \times S(\lambda) \times \Delta\lambda]$$

In broadband medium-pressure lamps (see Chapter 2), the effective germicidal power emitted in the range of 210 to 320 nm is about 50% of the total power emitted.

3.3 DOSE-EFFICIENCY CONCEPT

3.3.1 BASIC EQUATIONS

The basic expression of disinfection kinetics is a reaction of first order: $N_t = N_0 e^{-(k_1 t)}$ as long as the external parameters remain constant, k_1 in s^{-1}. On addition of a chemical disinfectant or irradiation (by intensity I), the reaction becomes one of apparent second order: $N_t = N_0 e^{-k_2 [It]}$, which is the Bunsen–Roscoe law indicating that under static conditions the disinfection level is related by a first-order equation to the exposure dose $[It]$:

$$N_t = N_0 \exp{-k[It]}$$

where

N_t and N_0 = volumetric concentration in germs after an exposure time t and before the exposure (time 0), respectively

k = first-order decay constant but depending on $[I]$

$[It]$ = dose, the irradiation power (in joule per square meter), also reported in milliwatt second per square centimeter). The SI expression of irradiation dose is joules per square meter, which equals 0.1 m Watt· s/cm^2. Various terms can be used for I: power, emitted intensity, radiant flux, or irradiance.

In theory, the active dose is the absorbed dose; however, as described in Section 3.2.3, the equations can be expressed on the basis of direct exposure dose. The latter represents the probability of efficient irradiation if appropriate correction factors for the relative efficiency at different wavelengths are applied (see, e.g., Table 7).

The basic kinetic equation is expressed in terms of dose (joule per square meter $[J/m^2]$), which stands for concentration as in disinfection by chemical oxidants. The potentially active dose needs to be evaluated according to the guidelines described and also as a function of the geometric factors as outlined in Section 3.7.

The decay law can be expressed as a Log 10 base as well as a log e basis; generally the Log 10 expression is used:

$$Log(N_t/N_0) = -k_{10}[It]$$

The D_{10} dose is the dose by which a tenfold reduction in bacterial count in a given volume is achieved. As long as the Bunsen–Roscoe law holds, this value can be multiplied to obtain the necessary dose for a desired log abatement (e.g., $4 \times D_{10}$ for a reduction by 4 log).

According to the logarithmic correlation between the remaining volumetric concentration of germs and the irradiation dose, the residual number of germs in a given volume can never be zero. Moreover, at high decay rates, discrepancies often occur in the log–linear relation between the volumetric concentration of germs and the irradiation dose. This effect can be described by assuming that for a given

TABLE 7
Numerical Values for the Potential Efficiency Coefficients at Different Wavelengths

λ nm	$S(\lambda)$	λ nm	$S(\lambda)$	λ nm	$S(\lambda)$
210	0.02	215	0.06	220	0.12
225	0.18	230	0.26	235	0.36
240	0.47	245	0.61	250	0.75
255	0.88	260	0.97	265	1.00
270	0.93	275	0.83	280	0.72
285	0.58	290	0.45	295	0.31
300	0.18	305	0.10	310	0.05
315	0	—	—	—	—

Note: The values are based on an approximation published by Meulemans [1986]. Cabaj et al. [2000] reported recently on the efficacy at lower wavelengths (see also Figure 50(b)). However, the principle of the approach remains unchanged.

bacterial population and strain, a limited number of organisms potentially resistant to disinfectants can exist in water: protected organisms N_p.

Accordingly, the Bunsen–Roscoe law can be reformulated [Scheible, 1985]:

$$N_t = N_o \exp(-k(It)) + N_p$$

By assuming that the number N_p is much smaller than N_o, the Bunsen–Roscoe law is still applicable for several decades of abatement.

3.3.2 METHODS OF DETERMINATION OF LETHAL DOSE

3.3.2.1 Collimator Method

One must use calibrated lamps of known emission spectrum. The mostly widely used method is given by the schematic in Figure 55.

The UV intensity is first measured and recorded. After this calibration, a bacterial suspension is placed in a cup having the same size as the window of a calibrated photocell operated in the cylindrical mode of detection. The cup is best made of strongly UV-absorbing material, to avoid reflections. The suspensions are exposed for variable time and the remaining bacterial numbers are counted after exposure and the data processed. The tests must be run at least in triplicate.

As for the photocells, they are mostly calibrated for the 254-nm wavelength. When using polychromatic sources, it is necessary to obtain information on the sensitivity of detection at other wavelengths and to integrate the whole, both sensor

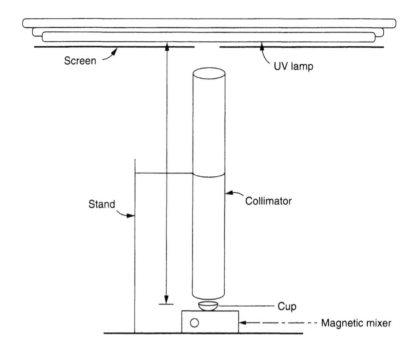

FIGURE 55 Setting up of a device for determination of D_{10} (laboratory collimator method).

detection rate and emission spectrum of the UV source again (e.g., by a 5-nm histogram approach).

The sensor detects and measures the incident intensity. For the real power (or flux) to be used in the dose computation, it may be assumed that about 4% of the power is lost by reflection at the free water surface. In other words, the power measured by the photocell must be reduced by 4% in the computation of the dose. If the water absorbs significantly in the UV range prospected, a correction factor for absorbance of extinction must be applied according to the Beer–Lambert law:

$$I = I_o \times 10^{-Ad} = I_o \times e^{-Ed}$$

where
I_o = blank measurement of the intensity
A and E = absorbance and extinction at different wavelengths, respectively
d = thickness of the liquid layer

Usually the thickness of the water layer is very small, so that this correction can be neglected. A more elaborate methodology for correction by competitive absorption is described in Section 3.7.2.

To operate such correction, the absorption spectrum of the water (or other liquid) must be known. As for the general absorption spectrum of drinking water, one can consider the loss of irradiation intensity of clear drinking water in a 5-nm segment histogram approach (λ as indicated ±2.5 nm), as shown in Table 8.

TABLE 8
Loss of Irradiation Intensity of Clear
Drinking Water in a 5-nm Segment
Histogram

λ (nm)	A (cm^{-1})	E (cm^{-1})	% Transmittance/cm
200	0.32	0.74	48
205	0.21	0.42	62
210	0.17	0.4	67
215	0.12	0.27	76
220	0.10	0.23	79
225	0.1	0.22	80
230	0.09	0.21	81
235	0.09	0.21	81
240	0.09	0.21	80
245	0.1	0.21	79
250	0.07	0.14	85
255	0.07	0.15	86
260	0.07	0.14	85
265	0.076	0.17	84
270	0.086	0.2	82
275	0.086	0.2	82
280	0.065	0.15	86
285	0.065	0.15	86
290	0.056	0.13	88
295	0.05	0.12	89
300	0.056	0.13	88

3.3.2.2 Correction for UV Exposure Cup Size

Often the cup of a liquid exposed to irradiation located under a collimated beam does not have the exact dimension of the collimated beam, nor the exact dimensions of the sensor. Therefore, geometric corrections are necessary. A recommended procedure is to measure the intensity as detected by the sensor in all horizontal X-Y directions at distances of 0.5 cm from the central focus of the beam. After summing all values thus recorded, divided by the number of measurements as well as by the value of the intensity recorded at the central focus point, one obtains a very average exposure intensity and consequently an exposure dose. (This correction often seems to be neglected in literature.) For further information see Tree et al. [1997].

3.3.2.3 Shallow-Bed Reactor

Shallow-bed, open-type reactors also can be used to establish reference doses [Havelaar et al., 1986]. Additionally, the technique is also more suitable for direct evaluation of the complete efficiency of medium (high)- pressure polychromatic sources, particularly when multilamp reactors are used. The reactor is shown schematically in Figure 56.

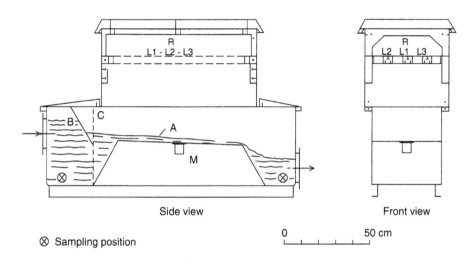

Side view   Front view

⊗ Sampling position

0   50 cm

FIGURE 56 Schematic of a shallow-bed reactor for lethal dose evaluations.

Water flows over a flat tilted bed (A), with the flow pattern streamlined and regulated by a baffle (B) and a perforated plate (C) with holes of 6-mm diameter. UV irradiation is produced by medium-pressure lamps: three in the case illustrated, Berson 2-kW lamps with a UV output of about 150 W (UV-C) per lamp and reflected to the water layer by an aluminum roof (R). Sampling points are (X) at the inlet and outlet zone (in option with automatic samplers equipped with refrigeration). Six quartz windows (M) are mounted in the irradiation bed (A) and measure the value of UV-C at these locations (used: MACAM type-three photometers equipped with a UV-C/P filter with cosine correction). Water depth is between 1 and 3 cm, depending on the water flow, which is kept between 10 and 30 m^3/h. The exact water depth is controlled by contact sensors. Blank standards are run with suprapure distilled water and, if necessary, the available intensity is corrected according to the Beer–Lambert law. (Because the water layer thickness is small, this correction stands for sewage and other absorbing liquids, instead of drinking water.)

3.3.3 Reported Values of D_{10}

Widely accepted values for D_{10} (in joule per square meter) are reported in Table 9. As for the total plate count that results from heterogeneous populations, a typical set of data is illustrated in Figure 57.

Claimed efficiencies of the Xenon-pulsed technology are at 300 J/m^2: 6-D_{10} for Enterobacteria, 2-D_{10} for enteroviruses, 4.3-D_{10} for *Cryptosporidium* oocysts; and at 400 J/m^2: 7.5-D_{10} for Enterobacteria, 2.6-D_{10} for Enteroviruses, and 4.6-D_{10} for *Cryptosporidium* oocysts [Lafrenz, 1999]. Long-term experience under real conditions still needs to be confirmed.

The dose required for algicidal treatment of water with UV is too high to be economically feasible and would require very large reactors when it comes to the treatment of large water flows. For these reasons and also other principles such as the potential

TABLE 9
1-D_{10} for Most Relevant Organisms Potentially Present in Drinking Water

Organism[a]	Value	Organism[a]	Value
Bacterium prodigiosus	7	E. coli (wild strains)	50[b]
Legionella pneumophila	9.2	Coliforms	50–60[b]
B. megaterium (vegetative)	11	Bacillus subtilis (spores)	300–400[b]
Streptococcus viridans	20	Bacterium coli	54
Yersinia enterocolitica	20	Pseudomonas aeruginosa	50–60[b]
(ATCC 23715)		P. aeruginosa	55
Legionella pneumophilia	20–50[b]	Infectious hepatitis virus A	58–80
Eberthella typhosa	21	(HVA)	
Shigella paradysenteriae	22	Somatic coliphages	60[b]
Dysentery bacilli	22	Streptococcus lactis	61
Streptococcus hemolyticus	22	Micrococcus candidus	63
Milk (Torula sphaerica)	23	Enterobacter cloacae	65
Serratia marcescens	25	(ATCC 13047)	
Salmonella typhi	25	Vibrio cholerae	66
(ATCC 19430)		Salmonella typhimurium	80
Escherichia coli	25	Enterococcus faecalis	80
(ATCC 11229)		(ATCC 19433)	
Klebsiella pneumoniae	25	Streptococcus faecalis	80[b]
(ATCC 4352)		S. faecalis (wild strains)	82
Proteus vulgaris	27	Rotavirus(es)	90
Bacterium megatherium	28	Adenovirus	300
(spores)		Bacillus subtilis (spores)	80–120
Citrobacter freundii	30–40	Micrococcus sphaeroïdes	100
Poliovirus	32–58	Clostridium perfringens	100–120
Rheovirus	110	(spores)	
Bacillus paratyphosus	32	Phagi f-2 (MS-2)	120[b]
Beer brewing yeasts	33	Chlorella vulgaris	140[b]
Corynebacterium diphteriae	34	(algae)	
Pseudomonas fluorescens	35	Actinomyces (wild strain	150–200
Baking yeast	39–60	spores Nocardia)	
S. enteritidis	40	Phagi f-2	240
Phytomonas tumefaciens	44	Fusarium	250–350[b]
Neisseria catarrhalis	44	Infectious pancreatic necrosis	600[b]
B. pyocyaneus	44	(virus)	
Spirillum rubrum	44	Tobacco mosaic virus	750[b]
B. anthracis	45	Giardia lamblia (cysts)	(400–800)[c]
Salmonella typhimurium	48	Lamblia-Jarroll (cysts)	700[b]
Aerobacter aeromonas	50[b]	L. muris (cysts)	700[b]
E. coli (wild strains)	50	Cryptosporidium oocysts[d]	7–10[c]

(continued)

TABLE 9
1-D$_{10}$ for Most Relevant Organisms Potentially Present
in Drinking Water (Continued)

Organism[a]	Value	Organism[a]	Value
Fungi spores	150–1000	Diatoms	3600–6000
Aspergillus niger	440–1320[b]	Green algae	3600–6000
Microanimals and parasites	1000 (?)	Blue-green algae (Cyanobacter)	3000

Note: The doses are expressed in joule per square meter, valid for suspensions of single organisms in pure water at pH = 7, at 22°C, in the absence of daylight, and in the linear part of the decay curve. In design, appropriate safety factors will need to be applied. The 1-D$_{10}$ doses indicated hereafter are the result of a large comparison and compilation of literature.

[a] No specific data seem to have been reported for nitrifying–denitrifying bacteria (*Nitrobacter, Nitrosomonas*). In case studies on wastewater treatment on a comparative basis, a nitrified effluent needs higher UV doses than a nonnitrified effluent.

[b] Specific data evaluated with medium pressure lamps.

[c] Data can be variable by a factor of 2 depending on the strain. Medium-pressure broadband emitting lamps can be more effective: 1-D$_{10}$ in the range of 400 to 800 for *Giardia lamblia* cysts and 7 to 10 for *Cryptosporidium* cysts. (In case of protozoan oocysts, the result can depend on the numbering method: excystation or *in vivo* testing.)

[d] For additional information: Bukhari et al., 1999; Clancy et al., 1998; Clancy et al., 2000; Clancy and Hargy, 2001; Hargy et al., 2000.

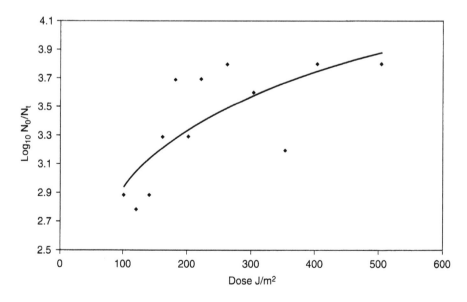

FIGURE 57 Decrease in total plate count (TPC) germs as a function of UV dose.

release of by-products on algicidal photolysis, the removal of algae and similar organisms has to rely on other processes currently used in water treatment.

Most of the data marked[b] in Table 9 are from Havelaar et al. [1986]. They concern measurements made with medium-pressure mercury lamps. It is comfortable to observe that the integration method in the UV-C range (see Section 3.2) gives equal results to those obtained with 254-nm low-pressure mercury lamps, except, however, in the case of bacteriophage f-2. Absorption by cellular proteins of part of the light emitted by medium-pressure lamps could be an explanation. At present, however, this hypothesis needs more investigation.

Little is known about the theoretical aspects of the killing effect of microorganisms and parasites with UV. However, the efficiency of broadband and multiwave lamps is well established in the field as far as *Cryptosporidium* oocysts are concerned (Figure 58).

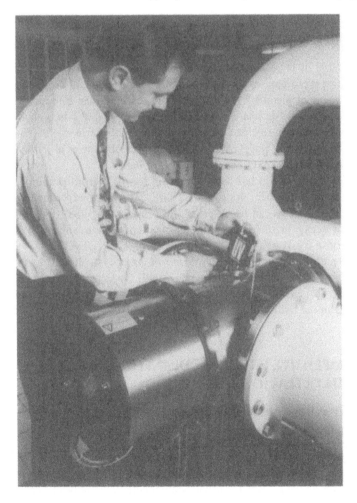

FIGURE 58 UV reactor of 8 Hg lamps of medium pressure emitting multiple UV waves for the elimination of *Aeromonas aerobacter*. (Berson installation at Culemberg [NL] 360 m^3/h at T_{10} = 78%.)

3.3.4 EFFECT OF WATER TEMPERATURE

The effect of the lamp temperature has been commented on in Chapter 2. The direct effect of the water temperature on the lethal dose for 22°C is negligible in drinking water treatment—less than 5 to 10% acceleration or slowing down, by either an increase or a decrease of 10°C [Meulemans, 1986].

3.3.5 EFFECT OF pH

The complementary effect of the pH of the water has not been investigated much. In experiments on distilled water, the pH generally has been maintained at 7. In investigations on drinking water, the pH was as such (i.e., between 7 and 8).

3.4 REPRESENTATIVE TEST ORGANISMS

From the table of D_{10} values, it can be considered that *Enterococcus faecalis* is a representative test organism for the group of Enterobacteria, and spores of *Clostridium perfringens* or phagi f-2 (MS-2) are more resistent than Enteroviruses. Spores often show a lethal-lag phase (see Section 3.6). Phagi f-2 is a more easy and representative criterion to check virucidal efficiency [Severin et al., 1984; Havelaar and Hogeboom, 1984; Havelaar et al., 1986; Masschelein et al., 1989]. See also Maier et al. [1995] and ISO-DIS 10705 [1993] Part 1.

A safety factor of 1.3 has been suggested for 4-D_{10} inactivation of viruses vs. the observed value for 4-D_{10} for phagi f-2 (MS-2). In some experimental conditions a biphasic decay curve can be observed, [Martiny et al., 1988] (tailing-off) after 2 to 3 logs of decay. In such a case an empirical correlation with the dose has been proposed: dose = a $[\text{Log}(N/N_o)]^2$ − b $\text{Log}(N/N_o)$ − c [Wright et al., 1999].

As for parasites, particularly *Cryptosporidium* oocysts, it appears that a lethal-tail phase also exists [Finch and Belosevic, 1999]. The investigations require highly concentrated suspensions of oocysts, which do not correspond to real concentrations of parasites in the field.

3.5 COMPETITIVE EFFECTS IN DISINFECTION WITH ULTRAVIOLET LIGHT

3.5.1 COMPETITIVE ABSORPTION BY COMPONENTS OF DRINKING WATER

The absorbance (log base 10) has been measured for the 254-nm Hg emission line. For evaluation in technical design, the transparency in percentage of a 10-cm layer is appropriate as well. Data for usual components potentially present in drinking water are listed in Table 10.

Multiwave lamps having a more diversified emission can remain active by the emissions that are less absorbed than at 254 nm.

TABLE 10
Absorbance at 254 nm of Potential Constituents
of Drinking Water

Constituent	A (in cm^{-1})	$\%T$ (1 cm^{-1})
Suprapure distilled water	10^{-6}	99.999...
Good quality groundwater	0.005–0.01	89–79
Good quality distribution water	0.02–0.11	63–78
Bicarbonate ion (315 mg/l)	35×10^{-6}	99.92
Carbonate ion (50 mg/l)	4×10^{-6}	99.99
Sulfate ion (120 mg/l)	48×10^{-6}	99.9
Nitrate ion (50 mg/l)	0.0025	99[a]
Fe^{3+} – $Fe(OH)_3$ (200 mg/l as Fe)	0.04	91
Aluminum hydroxide (hydrated 0.2 mg/l as Al)	Transparent at 254 nm	
Natural humic acids in water (according to Wuhrmann-Berichte EAWAG, Switzerland)	0.07–0.16	85–70
For comparative information		
Secondary clarified effluent	0.17–0.2	68–63
Groundwater with high-concentration humic acids[b]	0.11–0.5	78–32

[a] The absorbance of the nitrate ion and the possible formation of nitrite is discussed in more detail in Chapter 4. Humic acids can be a major optical interferent in the absorption of the 254-nm wavelength light. If present in natural sources, they are best removed before the application.
[b] See Eaton [1995].

3.5.2 STEERING PARAMETERS

From practical experience, the UV disinfection method requires specific evaluation in the design phase and special attention in operation if one of following parameters exceeds the very limiting values indicated:

Turbidity	>40 ppm SiO_2; or 16 NTU
Color	>10° Hazen
Iron content	>4 mg/l
BOD-5	>10 mg/l
Suspended solids	>15 mg/l
Amino acids and proteins	>3 mg/l

Turbidity often is the critical parameter considered. However, thanks to scattering of the light, the pathway is increased; and in some instances, turbidity can have a promotional effect on the disinfection efficiency [Masschelein et al., 1989]. In fact, general UV-C absorbance is an important overall parameter to be considered.

Note: Preformed chloramines do not lower the disinfection power of UV-C under conditions currently occurring in drinking water. In addition, under such conditions, no trihalomethanes (THMs) are formed in the presence or absence of monochloramine. Assimilable organic carbon (AOX) is not formed by application of UV alone, but can be formed when monochloramine preexists in the irradiated water [Blomberg et al., 2000]. Multiwave medium-pressure Hg lamps break down preexisting chloramines [B. Kalisvaart, private communication, 2001].

3.5.3 IMPORTANCE OF DISSOLVED COMPOUNDS

Dissolved iron in excess has a hindering effect, but has also been described to potentially exert a catalytic effect, the so-called the *NOFRE effect* [Dodin et al., 1971; Jepson, 1973]. The catalytic effect of iron during UV irradiation of algal extracts has been investigated more recently by Aklag et al. [1990]. However, it remains negligible at conventional dose rates.

The competitive effect of dissolved proteins has been described first by Mazoit et al. [1975]. All this information concerns low-pressure lamp technologies. Further evidence can be found in more recent investigations reported in Section 3.1 of this chapter [Aklag et al., 1990; Bernhardt et al., 1992].

The potential effect of some general organic compounds is illustrated by their absorption spectra, for example, as in Figure 59. Because good quality drinking water has an absorbance at 254 nm in the range of 0.02 to 0.11 (see Section 3.5.1), at less than 1 to 2 mg/L direct photochemical interference by organic compounds in

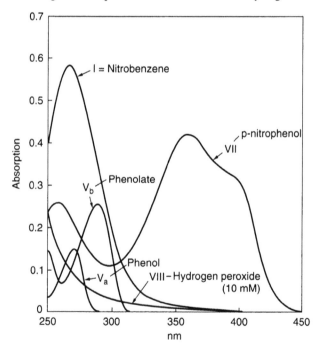

FIGURE 59 UV absorption spectra of some typical organic functions (according to Lipczynska-Kochany [1993]; absorbance per centimeter; Log base 10). The concentration of the organic compounds is 0.1 mM, for example, 10 to 15 mg/L. I, nitrobenzene; V_a, phenol; V_b, phenolate ion; VII, *p*-nitrophenol; VIII, hydrogen peroxide (10 mM).

disinfection of drinking water with UV light remains marginal, but not necessarily for photochemical-assisted oxidation processes (Chapter 4). Examples for absorbance at 254 nm (log base 10; in liter per mole and per centimeter) are 2610 for naphthalene and 10,000 for polychlorinated biphenyls (PCBs) [Glaze, 1993]. Hence, for example, PCBs at a concentration level of 2 mg/L dissolved carbon can represent an optical interference in disinfection efficiency of 254-nm UV corresponding to an additional absorbance of 0.025.

As a tentative conclusion with the present state of knowledge, competitive optical interference at a low concentration of organic micropollutants in drinking water remains of marginal importance in the disinfecting process with UV light. In photochemical oxidations the conclusion can be different (see Chapter 4).

Recently the bromate issue has been raised. The absorption of the hypobromite ion in the UV germicidal range is weak as long as submilligram per liter concentrations are concerned. As illustrated in Figure 60, the absorbance at submilligram per liter levels (concentration in Figure 60 is 0.15 mg/L), absorbance of the bromate ion is very small, so that direct photolysis of the ion in low concentrations in drinking water cannot be expected with conventional lamp technologies. Lamps emitting in the 200- to 220-nm range could have some efficiency (Figure 61; see also Figures 21, 22, and 27).

3.5.4 USE OF ARTIFICIAL OPTICAL INTERFERENCES IN INVESTIGATIONS

Parahydroxybenzoic acid has an absorption spectrum that matches the absorption of humic acids, and can be used as an internal optical competitive absorbent [Severin et al., 1984]. The absorbance depends also on the pH value of the water under investigation, as illustrated in Figure 62.

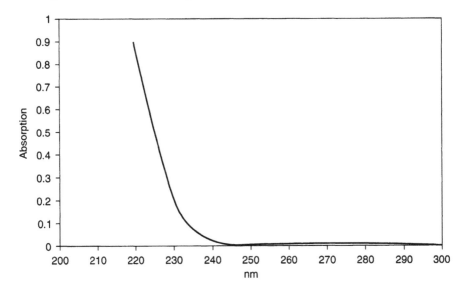

FIGURE 60 UV absorbance of bromate ion in water.

FIGURE 61 UV disinfection of process water in a brewery ($T_{10} = 95\%$ applied dose, 500 J/m^2), (Berson installation). (See also Figures 21, 22, and 27.)

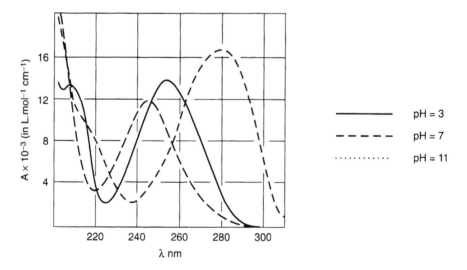

FIGURE 62 UV absorption spectrum of *p*-hydroxybenzoic acid.

Parahydroxybenzoic acid by itself has no bactericidal effect. At 10 mg/L with pH 7 and absorbance of 8000 cm^{-1} (at 254 nm), it enters into direct competition for the absorption of UV wavelengths. The method has been applied successfully in reactor modeling at 254 nm [Masschelein et al., 1989] (see Section 3.7). If used with polychromatic sources, again a correction by an histogram of the absorbance on the basis of 5-nm steps is necessary to evaluate the overall competition effect.

The use of fulvic acid, for example, isolated by the method described by Christman, is an alternative for optical masking [Severin et al., 1984].

3.6 MULTIHIT, MULTISITE, AND STEP-BY-STEP KILLING CONCEPTS

The experimental data often show discrepancies vs. the linear function of $Log(N_t/N_o) = -k[It]$ at low doses (i.e., at short irradiation time for a given technology), often there is a lethal-lag phase. From the technical point of view, the problem can be solved by providing an extra safety dose during design, as was done in research work on *Bacillus subtilis* spores [Qualls and Johnson, 1983]. The lethal-lag is sometimes considered as the result of partial photorepair after exposure to low doses [Bernhardt et al., 1996]. However, the phenomenon is more pronounced for multicellular organisms that cannot photorepair. A lethal-lag phase often also is observed in chemical disinfection—for literature on the subject see, for example, Masschelein et al. [1981]. More fundamental explanations are based on the multhit and multisite theories, as well as on the concept of consecutive reactions.

Assume that n "vital centers" each must be hit by an active photon to kill or inactivate the organism. Also assume a pseudo-first-order reaction for each center and photons in excess. If the first-order kinetic constant is equal for each center of a given type of organism (this is a reasonable hypothesis but certainly a weak point in the present state of fundamental knowledge), then with such preliminary assumptions one can express for the probability that n centers will be hit and the organism will be inactivated within the time t, as:

$$P_t = [1 - e^{-kt}]^n$$

The fraction of surviving organisms then becomes:

$$1 - P_t = [N_t/N_o] = 1 - [1 - e^{-kt}]^n$$

Using binomial extension of the probability of hit and killing and neglecting the term of a higher order than the first gives:

$$P_t = 1 - n e^{-kt}$$

and the decay rate becomes:

$$[N_t/N_o] = n e^{-kt}$$

or

$$Log[N_t/N_o] = -[kt/2.3] + Log\, n$$

By extrapolating the linear part of a plot of $log[N_t/N_o]$ vs. t to the origin, the ordinate at $t = 0$ corresponds to $Log\, n$. To study the phenomenon more closely at low exposure doses, the following (or a similar) experimental reactor may be recommended [Masschelein, 1986; Masschelein et al., 1989].

A low-intensity cold-cathode lamp light is used. The emission part of the lamp is submersible in water (e.g., the Philips TUV-6W(e) source). This is a monochromatic source (see Chapter 2) that merely emits at 254 nm, with the component at 185 nm eliminated by the optical glass of the lamp. The diameter of the lamp is 2.6 cm, the emissive length is 7 cm, and the UV (254-nm) intensity emitted is 0.085 W. The lamp is of instant start and also flash emissions can be produced, lasting between 0.5 and 10 sec by using a suitable timer (e.g., Schleicher-Mikrolais type KZT-11). A small correction of the irradiation time vs. the lightening time remains necessary at very short times (Figure 63). For hot-cathode lamps, the warmup time to obtain full regime is much longer. The lamps are best shielded during that period and the shield removed at time t_o.

The lamp is installed in a series of vessels with different diameters filled with seeded water and completely mixed (magnetic mixer). The exposure dose is corrected for the geometry factor m (see Section 3.7). A set of results is illustrated in Figure 64.

A very typical example is that of *Citrobacter freundii*. Both strains E-5 and E-10 studied converge to an n value of 3 (Figure 65). Most of the bacteria investigated show n values between 2 and 4, with the exception of *Proteus mirabilis*, which shows a rather speculative value of about 20, considering the lack of precision of the extrapolations in such a case. However, the value is high.

It is valuable for this approach to note that the values of n (i.e., 2 to 4 for bacteria in UV irradiation are similar to the ones observed in the lethal-lag phase investigations with chemical agents) [Masschelein et al., 1981, 1989].

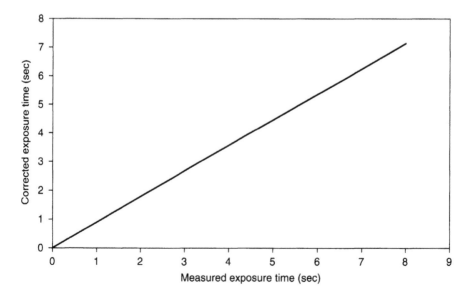

FIGURE 63 Correction of exposure time for instant start TUV-6W in water at 20 to 22°C.

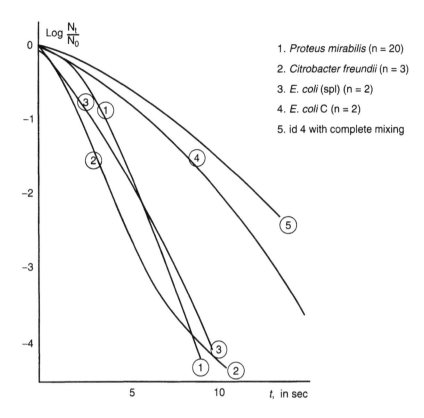

FIGURE 64 Experimental results concerning the lethal-lag phase. (From Masschelein, 1992, 1996.)

In the experiments with spores of *Bacillus subtilis* reported by Qualls and Johnson [1983], the Log n value was 1.01 or $n = 10$ (with a statistical confidence value of $r = 0.98$). This indicates that spores probably survive in water in the form of clusters.

According to the concept of multisite killing effect, different vital centers in a single organism are each to be hit once to be deactivated. The value of n is independent of the initial volumetric concentration of germs. The linear parts of the decay curves are parallel. In the multihit concept in which a given vital center must be hit several times before decay occurs, the linear parts of the decay graphs for different initial volumetric concentrations of germs are not parallel. This effect could be important in the inactivation of parasites as, for example, oocysts of *Cryptosporidium*. At a given period of multiplication, the parasite is indeed in the form of multicellular cysts. It is difficult, however, to clearly distinguish the two effects on the basis of experimental data.

Partially hit bacteria potentially also can repair after irradiation [Severin et al., 1984]. Therefore, it can be assumed that at least a minimum number of consecutive steps are necessary to achieve irreversible decay of a multicellular organism (and

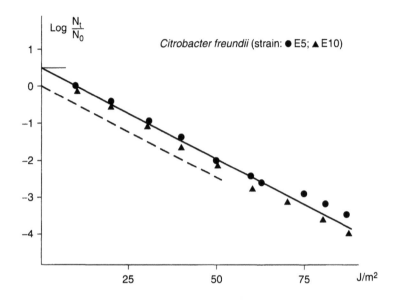

FIGURE 65 Normalized decay curve of *Citrobacter freundii* in water. (From Masschelein, 1992.)

even a monocellular organism):

$$B_o \rightarrow k_1 \rightarrow B1 \rightarrow k_1 \rightarrow B2 \rightarrow k_1 \rightarrow B(n-1) \rightarrow k_1 \rightarrow Bn$$

After n consecutive steps the decay occurs. At an intermediate stage, Bx, the change in volumetric concentration is given by:

$$\frac{N_x}{N_0} = \exp[-k(It)] \sum_{n=0}^{x-1} \frac{k(It)^x}{x!}$$

and the fraction of organisms surviving by:

$$\frac{N_x}{N_0} = \sum_{x=0}^{n=1} \frac{k(It)^x}{1+k(It)^{x+1}} = 1 - \left[1 + \frac{1}{k(It)}\right]^{-n}$$

In all preceding approaches, axial mixing (mixing orthogonally to the lamp axis) is assumed to be complete and water flow along the lamp axis is considered to be plug flow. All elementary first-order rate constants for the different steps are considered to be equal.

3.7 DESIGN FACTORS FOR REACTOR GEOMETRY

3.7.1 GENERAL

A point-source light is absorbed when irradiating a water layer. The generally considered absorption law is that of Beer–Lambert. On irradiating a layer of thickness d, the light intensity is decreased exponentially as a function of the layer thickness:

$$I_d = I_o \times 10^{-Ad}$$

or

$$I_d = I_o \times e^{-Ed}$$

the relative irradiation power becomes:

$$I_{rel} = I_d/I_o = 10^{-Ad} = e^{-Ed}$$

This approach allows quantification of the exposure to light in a water layer as in open channels, if the point-source concept is accepted. However, the latter has been called into question.

Channel-type reactors:

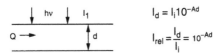

$$I_d = I_i 10^{-Ad}$$

$$I_{rel} = \frac{I_d}{I_i} = 10^{-Ad}$$

Cylindrical reactors (in to out):

$$I_r = I_i \frac{r_i}{r} 10^{-A(r-r_i)}$$

$$I_{rel} = \frac{I_r}{I_i} = \frac{r_i}{r} 10^{-A(r-r_i)}$$

Cylindrical reactors (out to in):

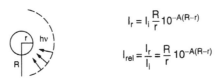

$$I_r = I_i \frac{R}{r} 10^{-A(R-r)}$$

$$I_{rel} = \frac{I_r}{I_i} = \frac{R}{r} 10^{-A(R-r)}$$

For perpendicular or orthogonal reactors, the same holds as for channel reactors. More details are to be found in comments on the aspect ratio of such reactors, Chapter 5, Section 5.4.

With cylindrical annular reactors and irradiation from inside to outside, the reactor walls are generally close to the walls of the lamp enclosures. The layer thickness considered is the difference between r_e and r_i. Accordingly, the relative intensity considered is expressed by:

$$I_{rel} = I_r/I_o = (r_i/r_e)\ 10^{-A(r_e-r_i)} = (r_i/r_e)\ e^{-E(r_e-r_i)}$$

For cylindrical reactors with irradiation "from outside to inside" the relations become:

$$I_{rel} = (I_r/I_o) = (R/r_i)\ 10^{-A(R-r_i)} = (R/r)\ e^{-E(R-r_i)}$$

3.7.2 Single-Lamp Reactors

An often used reactor configuration, especially for the treatment of low water flows (i.e., 5 m^3/h or less), is to locate a lamp with enclosure in the central axis of a reactor and to circulate the water in a void volume between the lamp enclosure and the walls of a cylindrical reactor. According to the Beer–Lambert law, the intensity lowers exponentially from the outside wall of the lamp (or enclosure) toward the inside wall of the reactor [Leuker and Hingst, 1992]. The effect is illustrated schematically in Figure 66.

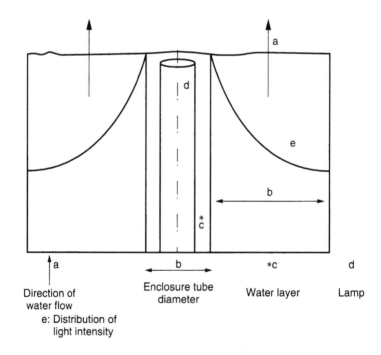

FIGURE 66 Schematic of irradiation profile in a single lamp cylindrical-type reactor.

An approximation for the effective dose in this type of reactor is to consider the irradiation dose at the wall of the reactor:

$$[It] = \text{exposure dose } (D) = L \times I \times T = I_o \times T \times L \times (r_e/r_i) \times 10^{-A(r_e-r_i)}$$

where
It = potentially biocidal UV dose (see Section 3.2)
S = maximum irradiation surface (m^2); A = $2\pi r_e L$; (L = length of the reactor)
T = irradiation time (seconds)
A = absorbance
L = length of cylindrical reactor
r_i = radius of the lamp + lamp enclosure
r_e = (internal) radius of the cylindrical reactor

This approach supposes a point-source in a completely mixed batch reactor. It is also implicitly assumed that in each segment of the lamp, the light is emitted orthogonal to the lamp wall. However, as the lamps emit in all azimuths according to their zonal distribution characteristics, part of the intensity is lost on the sides (S_1 and S_2) in the schematic shown in Figure 67.

The part of light that is lost is augmented by increasing R vs. L. A more complete approach is obtained by the conical model and the Lambert calculation [Hölzli, 1992]. The lamp is subdivided into a series of segments of equal length. Each segment is considered as an individual energy source Q, emitting an intensity I (Figure 68).

At a given point in the reactor, the intensity received from one given elemental segment then equals $I/4\pi d^2$; and by addition for the total number of segments one

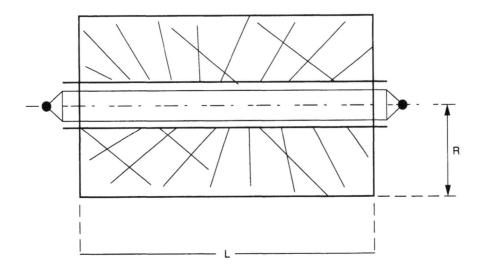

FIGURE 67 Schematic of a cylindrical reactor.

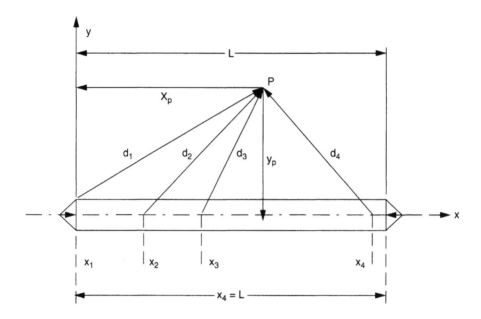

FIGURE 68 Schematic of the Lambert calculation model.

obtains the total intensity at a given point (P):

$$W_P = \frac{W_1}{4\pi d_1^2} + \frac{W_2}{4\pi d_1^2} + \cdots + \frac{W_n}{4\pi d_n^2} = \sum_1^x \frac{W_x}{4\pi d_x^2}$$

By integration over different points (y_P ordinates) the total intensity is obtained. The calculated values of I_P are in satisfactory agreement with the measured values with a photocell, at least for measurements in the gas phase.

Jacob and Dranoff [1970] used a similar model on a tubular reactor by including the presence of a quartz enclosure of the lamp in their approach. They found that a correction factor on the total intensity calculated according to the Lambert method is necessary to account for reflection and diffraction. This is accounted for by an empirical correction factor (C), depending on the location in the reactor space and given (by transposing to the symbols) as indicated in Figure 68.

$$C(y_P, x_P) = 1.0 + (y_P - 1.615)(0.13 + 0.0315x_P)$$

The experimental measurements at 350 nm on dilute solutions of 10^{-4} molar chloroplatinic acid, with correction for the sensitivity of the photocell and by the C factor, are within 2% in agreement with the calculated values (see Figure 69).

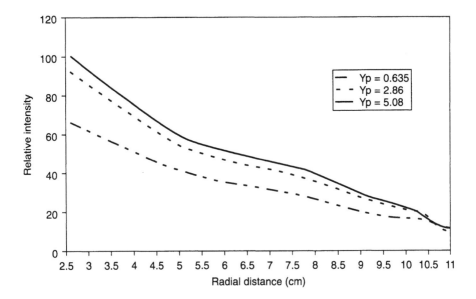

FIGURE 69 Light intensity distribution in an annular-space reactor. (Adapted from data of Jacob and Dranoff, 1970, with permission.)

One can still observe, however, that at shorter distances some deviations gradually occur between calculated and experimental values.

Qualls and Johnson [1983] published a modified version of the preceding approach, the so-called *point source summation method*. Availability of more powerful computer systems enables one to increase the number of point sources. They confirmed the model by bioassay using spores of *Bacillus subtilis* (ATCC 6633) in water. After correction for a lethal-lag phase, the accordance between the calculated exposure doses and the active doses for killing was acceptable. However, mixing conditions in the flow-through reactor were found to be important in bioassay methods (*vide infra*).

An attempt to account for the geometric distribution of light in an annular space around the lamp (or lamp enclosure) by considering that UV lamps are plasma emitters of finite dimensions (radius) has been formulated by Severin et al. [1984]:

$$(N_0 - N)Q = \int_0^l \int_0^{2\pi} \int_{r_i}^{r_e} \frac{kNI_0 r_i}{r} \exp[-E(r - r_i)] r \, dr \, d\theta \, dl$$

and:

$$(N_0 - N)Q = \frac{2\pi k NI_0 r_i l}{E} \{1 - \exp[-E(r_e - r_i)]\}$$

or:

$$\frac{N}{N_0} = \frac{1}{1 + mkI_0t} \quad \text{wherein} \quad t = \frac{V_v}{Q}$$

$$m = \frac{2r_i\{1 - \exp[-E(r_e - r_i)]\}}{E(r_e^2 - r_i^2)} = \frac{I_{rel}}{I_0}$$

The method defines a geometric correction factor m (always lower than unity for one single lamp). The relative light intensity I_{rel} equals mI_0.

Underlying hypotheses are that:

- The Bunsen–Roscoe law is applicable.
- The absorption obeys the Beer–Lambert law.
- The water flow along the lamp is plug flow.
- Axial mixing (i.e., mixing in units of void volume orthogonal to the axis of lamp and enclosure) is complete.

In such a case, for different values of extinction (E), standard curves can be established for different radii of lamp + enclosure as indicated in Figure 70.

Besides the data of the original work of Severin et al. (1984), a systematic investigation of the validity of the m-factor approach has been reported [Masschelein et al., 1989]. The bioassay is based on the decay of bacteriophage f-2 (which indicates no lethal-lag phase). A solution of p-hydroxybenzoic acid (PHBA) (10 mg/L), is used as an optical competitor in parallel experiences.

A cold-cathode Philips TUV-6W(e) monochromatic source with a lamp diameter of 2.6 mm (see Section 2.4.1.5) is installed in reaction vessels of different inner diameters (from 4.54 to 11.75 mm). The ratio of the UV (254-nm) power to the immersed surface of the lamp gives the value of $I_0 = 0.085$ W and $2 \times 1.3 \times 3.14 \times 7 \times 10^{-4} = 14.9$ W/m^2. The effective irradiation dose is given by D = mI_0t. A summation of data is given in Table 11.

The reaction vessels are filled with water seeded with bacteriophagi f-2, mixed by magnetic stirring, and flash-irradiated (see Sections 2.4.1.5 and 3.6, and Figure 63).

By applying the data to a 99% lethal dose effect (2-D$_{10}$), one obtains 473 $\pm$ 31 J/m^2 in water. The lethality being lowered in the presence of PHBA enables a comparison of 50% lethal dose rates and results in (joule per square meter), in pure water, 73 $\pm$ 6 and 68 $\pm$ 7 in water + PHBA. As a conclusion, the m-geometric factor correction is a valuable approach for the evaluation of irradiation doses in annular spaces around cylindrical lamps or lamp enclosures.

3.7.3 MULTIPLE-LAMP REACTORS

Single lamp reactors with an annular space (void volume) are operational in the water flow range of 1 to 10 m^3/h. Medium-pressure lamps, or high-rate low-pressure technology lamps can currently disinfect up to 50 to 100 m^3/h of water per lamp installed.

TABLE 11
UV Power Emitted by Submerged Portion of the Lamp

Φ (cm)	Void vol. (cm^3)	E (cm^{-1}) (water)	m	k_1 (s^{-1})	E (cm^{-1}) (water + PHBA)	m	k_1 (s^{-1})
4.54	125	0.046	0.712	−0.0429	1.84	0.340	−0.0255
6.49	300	0.046	0.546	−0.0370	1.84	0.155	−0.0107
8.42	560	0.046	0.442	−0.0295	1.84	0.087	−0.0055
10.3	850	0.046	0.369	−0.0221	1.84	0.057	−0.0033
11.75	1100	0.046	0.327	−0.0204	1.84	0.043	−0.0025

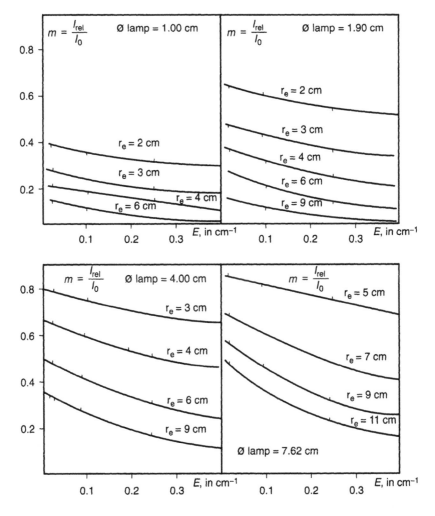

FIGURE 70 Geometric absorbance factors for annular space reactors. (According to Masschelein, 1992; 1996.)

Higher water flows per lamp become possible also by installing medium-pressure lamps in a transverse mode.

In the respective technologies, treatment of higher water flows necessitates multiplication of sequential single-lamp units or construction of multilamp reactors. In the multiple-lamp reactors, the configuration is called *positive*. In a single-lamp annular reactor the intensity levels decrease exponentially as a function of the distance between the lamp enclosure and the irradiated point considered (see Figure 66). In the multiple-lamp technologies, several lamps can irradiate a single point and thus create a local incremental effect. However, the shielding effect of one lamp by another must finally be taken into consideration as well.

A first approximation of the cumulative effect can be obtained by a simplified calculation: the lamps are supposed to be installed at equal distances along a circle of radius r in an annular space reactor of radius R, and the distance along the circle between the lamps is $2d$.

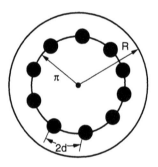

The number of lamps along the circle is:

$$n = 3.14 \times (R - d)/d$$

Suppose one wants to obtain a dose of $D = 250$ J/m^2 to treat a water flow of 175 m^3/h, or $Q = 0.05$ m^3/sec. If the reactor radius is $R = 0.6$ m and the useful reactor length is $L = 0.8$ m, the necessary exposure intensity is given by:

$$I = D\ (250\ \text{J/m}^2) \times Q\ (0.05\ \text{m}^3/\text{s})/3.14 \times R^2 \times L$$

$$I = 14\ \text{W/m}^2$$

From the irradiance curves of the manufacturers (see Chapter 2 and also an example in Figure 71), it can be noticed that the choice (if low-pressure Hg lamps are to be used) is the 30 W(e) lamp installed at distances between the lamps $2d =$

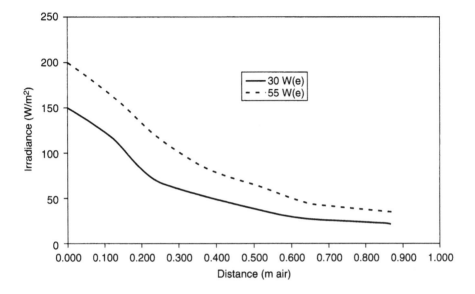

FIGURE 71 Typical irradiance curves of low-pressure Hg lamps. (Data taken from Osram: Technical Information Document MKAB/UV.)

2 × 0.14 m. (Another choice could be 55 W(e) lamps installed at 40-cm interlamp distance. This means higher operational costs.) The minimum number of lamps to be installed becomes:

$$n = 3.14 \times (0.6 - 0.14)/0.14 = 10$$

arranged at equal distance along a circle with a value of r of about $0.75R$, for example, 0.38 m, $d = 0.3$ m.

This first approach certainly is an oversimplification in design, but it can give a preliminary basis for design evaluations or first comparisons of existing lamp technologies:

- It supposes a single-point (single-line) source.
- It supposes no absorption by the water (air irradiance curves are considered).
- No cumulative effect on multilamp irradiation is taken into consideration.

A method based on the m geometric factor enables more accurate evaluation of the cumulative effect at a given point of a reactor volume equipped with several lamps by summing the different m values. In that way, iso-intensity reactors can be designed to realize the same exposure intensity at any point of the reactor. A typical example is given in Figure 72.

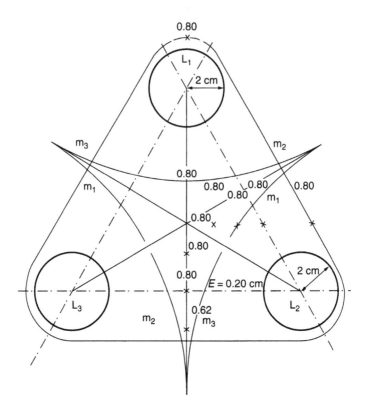

FIGURE 72 Triangular shape iso-intensity reactor. (Masschelein, W.J., 1992; 1996a,b.)

By adding the theoretical (i.e., by calculation, without considering reactor reflections on reactor walls) triangular modules into one single space, several arrangements can be computed. A classical arrangement is, for example, seven lamps located in a single cylindrical space (see Figure 73).

The intensity at any point in the reactor depends on the emission intensity of the lamp, the lamp enclosure radius, the distance between the lamps, and the extinction of the water (e.g., $E = 0.2$ cm^{-1}). A typical example of this technology is installed in Saalburg-Thüringen, Germany, to disinfect 4-Logs of *Escherichia coli* in a flow of 40 m^3/h [Leuker and Dittmar, 1992]. Such vessel-type reactors for the treatment of drinking water have been built to be equipped with 25 lamps (see Figure 74).

These types of reactors are equipped with lamps each having an electrical contact at one end and the other end is connected by wiring inside the quartz enclosure space. Lamps are standard 40 to 60 W(e). The maximum water flow (drinking water quality) that can be treated by one single unit is 400 m^3/h. The maximum working pressure is 16 bar (16 × 10^5 Pa). Similar experience has been made by KIWA at the Zevenbergen plant, the Netherlands. A typical six-lamp reactor installed for Gelsenwasser

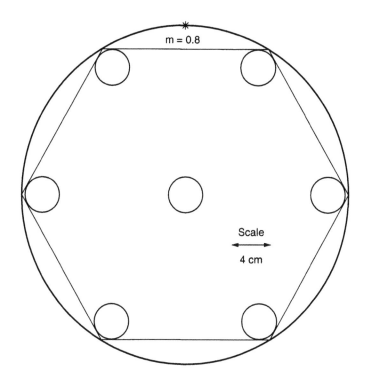

FIGURE 73 Schematic of a seven-lamp iso-intensity reactor.

(Höxter plant, Germany) is shown in Figure 75. With an extinction value of $E = 0.1$ cm^{-1}, this equipment can disinfect 180 m^3/h and realize an exposure dose of 360 J/m^2 at an electrical energy consumption of 1800 Wh(e).

Another type of reactor is based on the original doped flat lamp technology (see Chapter 2). The lamps are mounted in the water stream with the flat side perpendicular to the stream. The reactor is also shaped biconically to improve the reflection of light. A typical example is the five-lamp reactor (Wedeco design) installed for disinfection of drinking water at the city of Paderborn, Germany (see Figure 76). The electrical consumption is 770 Wh, and the reactor can treat up to 230 m^3/h at a germicidal dose of 350 J/m^2 at an extinction of $E = 0.04$ cm^{-1}.

Another example is the Berson reactor equipped with four medium-pressure lamps. The intensity distribution in the reactor volume has been studied particularly and optimized. The total germicidal intensity distribution is illustrated in Figure 77.

With water having an extinction of 0.22 cm^{-1}, a four-lamp unit can treat up to 400 m^3/h. Four clusters of units with the lamps mounted axially to the water flow of this type are installed, for example, at the waterworks of the city of Gouda, the Netherlands, to disinfect up to 1600 m^3/h. The medium-pressure technology is particularly well-suited for automation and remote control (Figure 78).

FIGURE 74 Multilamp vessel-type reactor.

FIGURE 75 Six-lamp reactor at Gelsenwasser, Germany. Installation Wedeco.

FIGURE 76 Reactor as installed at Paderborn, Germany (installation Wedeco).

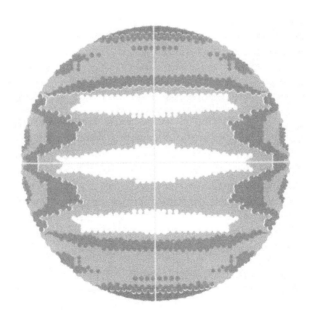

FIGURE 77 Geometric intensity distribution in a multilamp reactor (medium-pressure lamp Berson; UV intensity (watts): minimum 1630; maximum 3960; average 2850).

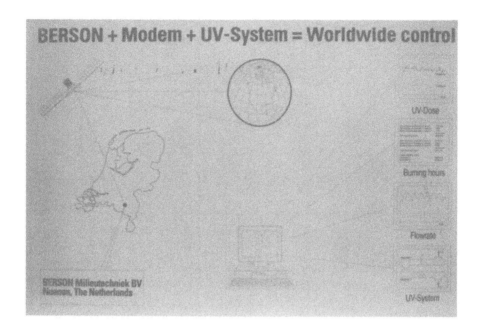

FIGURE 78 Berson-Modem worldwide control of UV systems.

3.8 MIXING CONDITIONS IN ULTRAVIOLET WATER TREATMENT

3.8.1 BASIC PRINCIPLES

For reactors in which the water flows along the axis of the lamps, ideally the flow-through pattern of the UV reactors for disinfection should be plug flow along the axis of the lamps and completely mixed in the space orthogonal to the lamps [Thampi and Sorber, 1987]. The same holds for reactors in which the lamps are cross-mounted vs. the water stream.

The theoretical basis for the mixing conditions and their impact on the decay rate have been defined explored and reported by Severin et al. [1984]. Based on pseudo-first-order decay rate constants one has for perfect plug flow:

$$\frac{N_t}{N_0} = \exp[-k(It)]\sum_{x=0}^{n=1}\frac{k(It)^x}{x!}$$

and, for completely mixed reactors:

$$\frac{N_t}{N_0} = 1 - \left(1 + \frac{1}{kIt}\right)^{-n}$$

A more complete model has been developed by the Cyclone firm in the Netherlands, supported by Berson.

3.8.2 GENERAL HYDRAULIC CONDITIONS

Hydraulic conditions have been studied extensively [Scheible et al., 1985] in the case of wastewater treatment. In all instances, the flow must be turbulent (Reynolds number: Re ≥ 2000). This is determinant in wastewater treatment, but it also holds for drinking water treatment. However, the concept should not lead to constructions with a very narrow annular space between the lamp enclosure wall and the reactor wall (to avoid losses), because a large number of photons could remain unabsorbed. A compromise often is necessary.

Various devices have been used to promote appropriate mixing in UV reactors: incorporating baffles, static mixers, or conical elements in the reactor pipe [Cortelyou et al., 1954]; placing the sources in the turbulent area of an ejector [Aklag et al., 1986]; and using automatic wiping devices. The latter also can serve for continuous cleaning of the lamp enclosures. Naturally, the construction must be such that no significant part of the light is shielded by the mixing elements. Finally, the hydraulic design of the reactor inlet–outlet is important and construction-specific.

3.8.3 TESTING OF FLOW-THROUGH PATTERNS

At the commissioning phase, the flow-through pattern of a reactor should be tested experimentally *at nominal, maximum possible, and minimum expected flow regimes.* The common method is that of chemical tracing by injection of a product, resistant to the UV irradiation and easy to measure at the outlet. Injection of a concentrated salt solution and continuous measurement of the electrical conductivity is a simple method [Thampi and Sorber, 1987]. The principle is illustrated in Figure 79.

The average velocity of the water (v) is given by L (length) × Q (flow) divided by the void volume V:

$$v = (L \times Q)/V$$

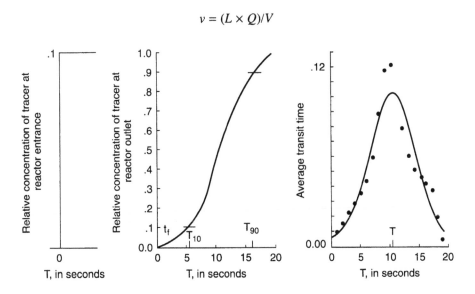

FIGURE 79 Testing of reactor flow-through pattern.

The variance in residence time at the reactor outlet $(\sigma_t)^2$ defines a dispersion coefficient D:

$$D = 0.5 \times v \times L \times (\sigma_t)^2/T^2$$

(T is the observed average residence time.) The lower the dispersion coefficient, the better it is for disinfection efficiency.

The Morrill index equals the ratio of times for 90 to 10% of the tracer to pass. In perfect plug flow, T_{90}/T_{10} approaches 1. In UV systems, the recommended value is to be lower than 2 at all flow-through regimes.

Qualls and Johnson [1983] have used spores of *Bacillus subtillis* as a tracer instead of a chemical. In the system they investigated, the Morrill index was 1.7. The bioassay method has the advantage of also controlling the disinfection efficiency and at the same time the hydraulic characteristics of the reactors; however, the method is more labor intensive. At present, bacteriophage is the preferred trace organism. Spores of *B. subtilis* are an alternative [Sommer et al., 1997].

3.8.4 LONGITUDINAL OR TRANSVERSE MOUNTING OF LAMPS

Berson intensively studied the flow-through patterns of lamp installations in reactors, particularly at turbulent regimes. In a conventional longitudinal arrangement, potential hydraulic short-circuiting can occur, but it can be kept under appropriate control by design, as illustrated in Figure 80.

By mounting the lamps in the traverse mode, more uniform distribution of UV intensity is obtained, as well as less formation of deposits on the lamps (and lamp enclosures). In addition, the distribution of the water is approximately steady over the entire reactor section, as illustrated in Figure 81. The company Berson-UV developed extensive computer modeling of both the hydraulics of the reactors and the distribution of UV intensity of the lamps installed. It is now practically possible to "tailor-to-measure" such equipment.

Figure 82 illustrates a still more recent arrangement, not only for disinfection purposes but also for appropriate synergistic oxidation with UV and hydrogen peroxide.

3.9 OPERATIONAL CONTROL OF EFFICIENCY

3.9.1 DIRECT CONTROL

Direct control of the disinfection efficiency of the installed systems is based on comparative microbiological counts of the incoming and outgoing water. The most suitable test organisms are the total plate counts (TPC) at 22 and 37°C, coliform and *E. coli* counts, and occasionally (at monthly periods) spores of *Clostridium perfringens*. Coliphage f-2 (also termed as a variant MS-2) is a significant organism for control. Personally, we also find the occasional control of spores of *Actinomyces* (even with a simplified procedure—Masschelein [1966]) appears relevant in relation

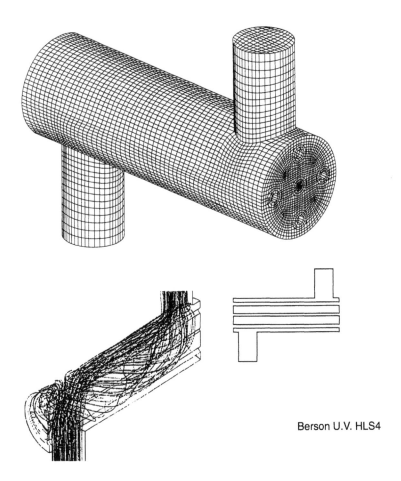

Berson U.V. HLS4

FIGURE 80 Hydraulic flow-through profiles in longitudinal reactors.

to the probability of the presence of viruses. Additional controls for commissioning the installation are specified in Section 3.10.

3.9.2 PERMANENT MONITORING

It must be recommended to install one or more photocells (depending on the size of the reactor) either at the outer wall or inside the reactor, which evaluate the UV output of the lamps continuously by visual reading of the records and by having alarm levels or automation installed depending on the size of the equipment.

This record does not necessarily have to be a precise, accurate, and absolute measurement in terms of physics and photochemistry (see Chapter 2), but it needs to be reliable in indicating on a relative basis at least any drop in efficiency of emission intensity and of (accordingly) overall irradiance into the reactor.

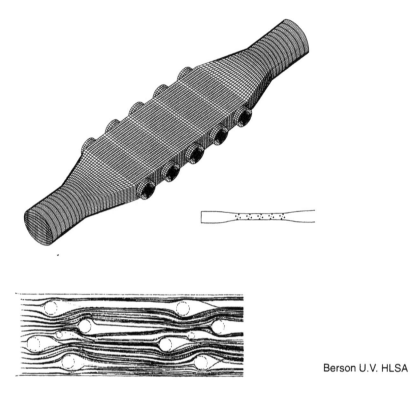

Berson U.V. HLSA

FIGURE 81 Typical axial distribution in a Berson UV reactor.

In a side detector of the reactor (see the following schematic) the relative positions of the detector and the lamp can be important and possibly detect mainly one of the emission sources.

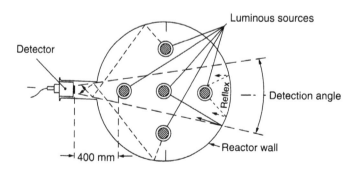

When locating the sensor outside of the reactor, one must take into consideration a loss of intensity by geometric factors also. In a typical outside location of the detector (see the following schematic), correction factors for real detection are necessary.

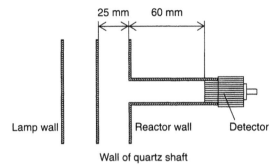

For example, if the light transmission of the water (10 cm) is 65% overall, then at 2.5 cm it will be 90% of the total; at 6 cm, 76% of the total; and at 8.5 cm (6 + 2.5) cm, 59.4% of the total. Fluctuations in intensity vs. the real intensity (I_0) will be indicated vs. 40% of the absolute value. As a conclusion, only part of the intensity is detected and recorded, but relative fluctuations can be used as a monitoring control.

Reduced scale (mini) photocells are available at present and can be located inside the reactor to average the measurement of more real exposure intensity (see the following schematic).

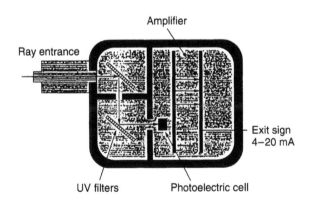

Standardized equipment is available to calibrate or recalibrate the cell of a given monitor with a calibrated unit (see Chapter 2). Some monitors are equipped with appropriate filters and collimators to match as closely as possible the general curve of microbial disinfection action of UV. Some of such photocells try to match the sensitivity of detection at different wavelengths as closely as possible to the germicidal wavelength. An example is illustrated in Figure 83, according to the prospectus of Hamatsu in Japan.

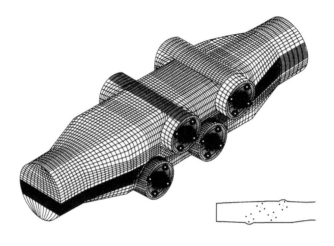

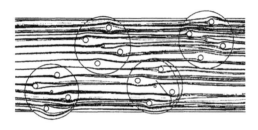

FIGURE 82 Image of intensity distribution in a high-yield reactor multiple-lamp traverse mode (16 medium-pressure lamps per clusters of 4 in the alternating traverse mode).

Again, a 5-nm histogram approach is recommended, superposing the emissivity of the lamp between 200 and 300 nm vs. the relative detection sensitivity (taking, by tradition or conservatism, the unit at 254 nm).

3.9.3 EXTENDED CONTROL

At least once a year, at the occasion of maintenance, a full control and checkup of the system are recommended. The minimum checklist should contain:

- Checking (and if necessary replacing) the lamps
- Cleaning and controlling the UV transparency of the lamp enclosure materials
- In case of doubt, checking the overall UV emission efficiency (lamp + enclosure)
- *Cleaning and checking the reliability of the controlling photocells after aging*
- Proceeding to general mechanical maintenance of the system

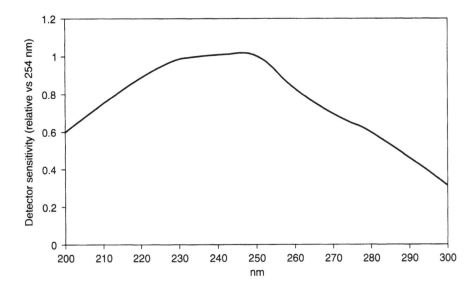

FIGURE 83 Matching of response of a UV monitor vs. disinfection. (According to documents from Hamatsu, Japan.)

3.10 TENTATIVE DESIGN QUESTIONNAIRE FOR ULTRAVIOLET-BASED DISINFECTION UNITS FOR DRINKING WATER

3.10.1 GENERAL

At present only few standards apply to the design of UV disinfection of drinking water: Austria Önorm—1993, updated in 1996 and again in 2001; Bavaria—1982; FIGAWA—1985. These texts recommend a minimum exposure dose of 250 J/m^2 based on the 254-nm wavelength. Actual rules in Austria are 400 J/m^2.

Further essential parameters for evaluation are [Meulemans, 1986; Cabaj et al., 2000]:

- Emission spectrum of the UV source
- Expected lifetime of the UV source and aging characteristics
- Temperature dependence of the UV source
- Optical absorbance (or transmission rate) of the water in the active UV range (i.e., at least from 210 to 320 nm)
- Allowable survival ratio and selected test organisms for control of efficiency

More detailed general information involves:

- Method of monitoring (not necessarily measurement of) constant UV irradiation
- Average water residence time in the reactor zone
- Applicable geometric factor (m) to apply (see Section 3.7.2 of this chapter)

3.10.2 Definition of Performance Required

To obtain the most appropriate offer and the best guarantee of success, the problem to be solved must be defined as accurately as possible. This implies at least:

- Origin of the water to be treated (source, lake, groundwater, surface water, etc.)
- Pretreatment, if any
- Ionic balance (with eventual seasonal minimum–maximum values) of the water to be treated
- *Possible seasonal variation in water temperature* and pH values
- Possible content of iron and manganese
- Possible turbidity and amount of suspended solids (milligrams per liter)
- Possible total and dissolved organic carbon
- Information on specific pollutants (if any significant)
- Total plate count (TPC)
- Coliform and *E. coli* counts
- If available, counts for *Clostridium perfringens*, bacteriophages, and specific parasites
- Definition of treatment objectives, microbiological standards to be met
- Commissioning parameters
  - Hydraulic tests at various flow regimes
  - Bioassay methods
  - Control methods of disinfection performance
  - Measurement of power consumption

3.10.3 Elements of Qualification and Tendering

3.10.3.1 General Presentation of the Offer

- Manufacturer (of reactor and of lamps and quartz enclosures, with address, phone, fax, e-mail, and name of contact person)
- Possible flow to be treated, cubic meters per second (or per hour)
- UV absorbance (full spectrum 210 to 320 nm), per centimeter or per meter
- Type of reactor, materials, schematic, etc.
- Nominal flow at 100% transmittance, cubic meters per hour
- Nominal flow at absorbance indicated earlier, cubic meters per hour
- Admissible water pressure, pascal
- Headloss at nominal operation, pascal
- Admissible variations in water flow, cubic meters per hour
- Feed current volt and hertz
- Type of electrical connection, standard applied
- Total electrical consumption, watt-ampere
- Number of lamps
- Power consumption of one lamp, watt
- Total UV yield of one lamp (+ spectrum), watt

- Absorbance of enclosure material (+ spectrum), per centimeter or per meter
- Expected lifetime of lamps and enclosure material, hour
- Principle of mixing (if any)
- Principle of continuous monitoring
- Degree of automation
- Cleaning device and mode of operation
- Conditions of water feed (hammer protection, etc.)
- Maintenance requirements and recommendations
- Possibility of maintenance contract
- Recommended number and type of spare parts
- Start–stop procedures
- Maintenance conditions during periods of out of service (long and short)

3.10.3.2 Operational Guarantee

- Financial guarantee
- Guarantee on hardware
- Guarantee on components (if different from hardware)
- Recommended insurances
- Obligations to be met by the user

3.10.3.3 Cost Parameters

- Reactor hardware cost, per unit
- Cost of lamps, per unit
- Cost of lamp enclosure, per unit
- Cost of UV monitoring device, per unit
- Cost of hydraulic protection of the system, per reactor or subunit
- Cost of automation (if any)
- Cost of mounting on site
- Cost of assistance on starting-up
- Costs and conditions of a preliminary pilot investigation (on a separate basis from the main offer)

3.10.3.4 References

- To similar installations in operation
- To standards applicable
- To evaluations of similar equipment made by official bodies
- To certifications, ISO, CEN, EPA (if any are applicable)
- To specific descriptions in literature and scientific publications

Note: Existing standards may not be a hindrance for innovation and progress.

3.10.3.5 Other Aspects

The tender may include other aspects more specifically.

3.11 EXAMPLES

3.11.1 SOURCE DU PAVILLON AT SPONTIN, BELGIUM

The first installation for disinfection of drinking water by a Belgian utility was equipped in 1958 for the Source du Pavillon at Spontin (Figure 84). The water is mixed *in toto* with that of other winnings, and after preventive disinfection with chloramines it is transported to Brussels by gravity through an aqueduct.

The neighboring municipality of Sovet has its own water well and a distribution system with a service reservoir, but the available quantity is not always sufficient. A supplement was ensured by pumping up to 5 m^3/h from the clearwell of the Source du Pavillon (i.e., upstream the *in toto* disinfection). Therefore, a UV unit was installed. The unit concerns a very classical single low-pressure lamp reactor, lamp + enclosure diameter (2 cm), and water layer thickness 1 inch. The reactor housing was made of galvanized steel. The system operated well and was replaced in 1981 by a new construction of stainless steel equipped with three lamps, 40 W(e) each, in series able to treat 15 m^3/h in total. After 23 years of operation, the first equipment

FIGURE 84 Photo of the source du pavillon at Spontin.

installed in a permanently moist atmosphere showed some early signs of corrosion (see Figure 85). For requirements of space, the reactors were fixed at the roof of the pavillon.

Practical experience gained from the application:

- Besides the yearly replacement of the lamp, it is necessary to clean the lamp enclosure (in this case at least twice a year).
- The quartz lamp enclosure surrounding the lamp should be made accessible without having to disassemble the whole unit.
- The (quartz) lamp enclosure gradually loses its transparency (at least at ≤254 nm). Transparency after 3 years became less than 60%, so the quartz enclosure was replaced. The phenomenon of aging of the quartz enclosures (perhaps by solarization) seems to be less studied in practice.

In the new design (1981) (Figure 86):

FIGURE 85 UV reactor installed in Pavillon at Spontin in 1958.

FIGURE 86 Updated UV reactor installed in Pavillon at Spontin in 1982.

- Results were always satisfactory for coliforms as they were for *E. coli.*
- At lamp aging, reduction in performance for TPC bacterial number is observed and aftergrowth after storage of 24 h in the laboratory occurs. This is observed before break-through of enterobacteria, and is a safe signal to take action.
- In the application, the off-take of water is variable with time. In case of no off-take, a minimum water circulation through the reactors is maintained (0.2 to 0.3 m^3/h) in a closed loop to maintain the system.

3.11.2 IMPERIA, ITALY

The Imperia facility (installed by Berson, the Netherlands) (Figure 87) produces 1200 m^3/h of drinking water by disinfection of groundwater with low-pressure Hg lamps. Four reactors are installed in series, each equipped with 12 lamps of 80 W(e).

3.11.3 ZWIJNDRECHT, THE NETHERLANDS

The Zwijndrecht (installed by Berson, the Netherlands) (Figure 88) facility disinfects riverbank-infiltrated water with T_{10} = 85%. Four units are each equipped with four medium-pressure multiwave UV lamps.

3.11.4 ROOSTEREN, THE NETHERLANDS

The Roosteren facility (installed by Berson, the Netherlands) (Figure 89) treats 1000 to 1600 m^3/h by UV disinfection of groundwater (T_{10} 97%). Four reactors operate in parallel, a total of 20 multiwave medium-pressure lamps.

FIGURE 87 Imperia, Italy-AMAT: 1200 m^3/h, disinfection of groundwater with low-pressure Hg lamps; four reactors in series, each equipped with 12 lamps of 80 W(e).

FIGURE 88 Zwijndrecht, the Netherlands: disinfection of potable water from bank-filtered water; $T_{10} = 85\%$; four units of four Hg medium-pressure lamps each emitting multiple UV wavelengths.

FIGURE 89 Roosteren, the Netherlands: 1000 to 1600 m^3/h at T_{10} 97% for disinfection of groundwater; 20 Berson 2500 (medium-pressure Hg with multiple emission wavelengths) installed in four parallel reactors.

3.11.5 Méry-sur-Oise, France

At the large Méry-sur-Oise (installed by Berson) (Figure 90) facility, 7480 m^3/h of surface water is treated conventionally followed by nanofiltration; T_{10} 90%. Five units in parallel are each equipped with four medium-pressure multiwave (B 2020) lamps.

FIGURE 90 Méry-sur-Oise, Paris, France: surface water previously treated and filtered through membranes. Normal output 7480 m^3/h at T_{10} = 90% by five units in parallel, each with four B2020 medium-pressure Hg lamps (emitting multiple wavelengths).

4 Use of Ultraviolet in Photochemical Synergistic Oxidation Processes in Water Sanitation

4.1 BASIC PRINCIPLES

4.1.1 GENERAL

Photochemical synergistic oxidation processes are a recent development in water treatment, related to the necessary removal of pollutants that are resistant to the more classical methods of treatment. The techniques, still in further development, are often termed commercially advanced oxidation processes (AOPs).

Besides the chemistry specifically related to ozone (for an overview, see Hoigné [1998]), these technologies involve several aspects related to the application of ultraviolet (UV):

- Direct photolytic action on compounds dissolved in the water sources
- Photochemically assisted production of oxidants (mainly supposed to be hydroxyl free radicals)
- Photochemically assisted catalytic processes

Although effects have been observed on the ground, it must not be forgotten that an overall energetic balance is required.

Considerable amounts of data have been reported in the literature related to water treatment, both in laboratory experiences, pilot plant investigations, and full-scale applications. However, even when the conditions and methods applied have been described with precision, it is often not possible to formulate general guidelines for design from the positive evidence as reported. These oxidation methods are applicable for the removal of compounds resisting the more classical techniques. This effect is often considered as secondary in technical literature. More investigation on this subject is required. It certainly plays a role in combined ozone-UV processes [Denis et al 1992; Masschelein, 1999; Leitzke and Friedrich, 1998].

The aim of this chapter is to summarize some fundamental aspects of these applications and to tentatively indicate preliminary recommendations for future design rules and experimental protocols to be formulated and to apply.

A fundamental characteristic of UV light is that the photons of these wavelengths are of sufficient energy to raise atoms or molecules to excited electronic states that are unstable in environmental conditions. These tend to transfer energy either by returning to the ground state or by promoting chemical reactions. Typical UV absorbance domains of a number of organic compounds are given in Figure 91.

The excited electronic state can be the result of either an *ionization* or an *activation* of the irradiated molecule or atom. Ionization can be represented as:

$$M + h\nu = M^+ + e^-$$

The electrons produced that way can either promote photoelectric processes or act as reducing agents: $C^+ + e^- = C$.

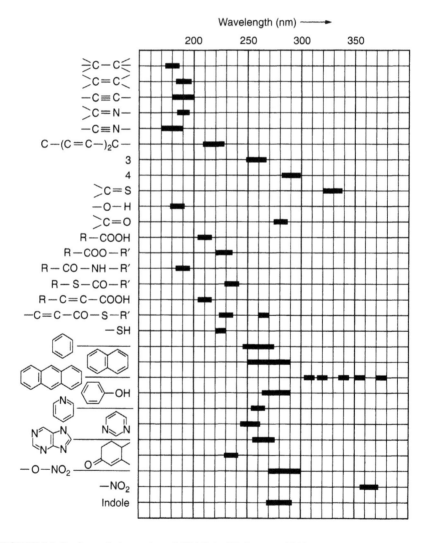

FIGURE 91 Regions of absorption of UV light [Kalisvaart, 2000].

Activation can be shown as:

$$M + h\nu = M^*$$

Several mechanisms of deactivation of M^* can occur:

- Thermal dissipation (which is not interesting for water treatment)
- Photonic energy transfers, as by fluorescence, that is, energy transfer to other molecules or atoms of lower energy state of activation (e.g., chain reaction mechanisms)
- Rupture of linkages between atoms in molecules

The two latter mechanisms can be significant in water treatment.

The direct effect of the 253.7-nm wavelength of the low-pressure mercury lamps on the decomposition of dissolved chlorinated hydrocarbons has been studied as early as 1986 [Frischerz, 1986; Schöller, 1989]. To obtain removal of trichloroethene and trichloroethanes by 40 to 85% in conditions of germicidal treatment, an irradiation time of 1 h was required.

Sundstrom et al. [1986] reported the direct photolysis of halogenated hydrocarbons. For example, 80% removal of trichloroethylene from a solution at 58-ppm concentration needs an irradiation time of 40 min. Other experiments similarly concern the irradiation of chlorinated aromatic compounds. Weir et al. [1987] reported similar yields for the abatement of benzene. Zeff and Leitis [1989] patented results on direct photolysis of methylene chloride. With conventional equipment, an irradiation time of 25 min was required to obtain an abatement of ca. 60% when starting from solutions of 100-ppm concentration.

Guittonneau et al. [1988] studied the oxidation of THMs and related halogenated ethanes in a batch reactor system. The conclusion was that evaporation losses may not be neglected in the experimental conditions as applied and that no evidence could be produced in the experiments for the rupture of C–Cl bonds. Nicole et al. [1991] investigated again the potential destruction of THMs in annular reactors. They found that C–Br bonds can be photolyzed, but only after long exposure times (e.g., 30 min or longer).

The UV-B range also has been prospected and may be important for the application of medium-pressure Hg lamps. Dulin et al. [1986] reported on the photolysis of chloroaromatic compounds in water by irradiation with medium-pressure mercury lamps from which the UV-C was removed by optical filters. Simmons and Zepp [1986] found that at 366 nm, humic substances could produce an inner filter effect (which is optical competition by absorption of at least part of the light), on the photolysis of nitroaromatic compounds. Peterson et al. [1990] studied the direct photochemical degradation of pesticides in water with a medium-pressure mercury lamp. Toy et al. [1990] prospected Xenon-doped arcs to remove 1,1,1-trichloroethylene. Up to 80% removal could be obtained after 30 min of irradiation. Finally, Eliasson and Kogelschatz [1989] have developed excimer sources capable of ionizing or activating C–Cl bonds more specifically. This development is still in an experimental stage as far as drinking water treatment is concerned.

It can be concluded that direct photochemical reactions with trace concentrations of organic micropollutants are of low efficiency and would require high irradiation doses to be operated. Reaction times mentioned by the authors range between 25 and 60 min with germicidal lamps. By comparison, average hydraulic residence times in UV disinfection units are in the range between 1 and 15 sec. This means that direct photooxidation would require UV dosages in the range of 40,000 to 80,000 J/m^2.

The possible reactions, however, may not be neglected as potential secondary effects in the synergistic oxidation processes. Most of the principles of photochemically assisted oxidations in water treatment are, at the present state of knowledge, considered as •OH-radical chemistry.

Direct photooxidation of water is important in photosynthesis [Rabinowitch, 1945]. Under conditions of water treatment, however, vacuum UV light is required to directly dissociate water into reactive H• and •OH radicals. Another method is based on photocatalytic processes, as discussed in Section 4.4. In the synergistic oxidation processes, •OH radicals also are produced by photolysis of either ozone or hydrogen peroxide.

Vacuum UV, xenon excimer lamps (172 nm) are in full development [Eliasson and Kogelschatz, 1989] for the direct production of radicals on irradiation of water. Applications for general water treatment are not yet expected considering the limited size of the equipment and the yet undefined cost.

4.1.2 CHARACTERISTICS OF •OH RADICALS RELATED TO WATER TREATMENT

Hydroxyl radicals have both oxidation and reduction properties. The standard redox potential (i.e., vs. normal hydrogen electrode, calculated) of •OH is 2.47 V (values up to 2.8 V are published). The reducing properties are, as suggested by Weiss [1951] due to dissociation: $•OH = O^- + H^+$. The reducing properties have been attributed to the oxygen mono-ion. Furthermore, the reducing properties of •OH can determine back-reactions in oxidations of ions as, for example:

$$Fe^{2+} + •OH = (Fe^{3+} - OH^-)$$

followed by

$$(Fe^{3+} - OH^-) + •OH = Fe^{2+} + H_2O_2$$

In the case of iron salts, the first reaction is the most important, but with other polyvalent ions (e.g., cerium salts), the reduction pathway can become more important [Uri, 1952]. These types of reactions have not yet been considered exhaustively in water treatment, and at present the oxidation pathway is most described.

The O–H bond dissociation energy is estimated as (418 ± 8) kJ/mol [Dwyer and Oldenberg, 1944]. The overall energetic aspects of reactions of •OH radicals and related oxygen species *in the aqueous phase* are reported according to Uri [1952] (data in kilojoule per mole):

$\bullet OH + H_2O_2 = H_2O + HO_2$ (radical)	79.5
$\bullet OH + HO_2$ (radical) $= H_2O + O_2$	322
HO_2 (radical) $+ H_2O_2 = H_2O + \bullet OH + O_2$	125.5
HO_2 (radical) $+ HO_2$ (radical) $= H_2O_2 + O_2$	242.7
$\bullet OH + \bullet OH = H_2O_2$	196.6
$\bullet OH + \bullet OH = H_2O + O$	62.8

Halogen ions inhibit the reactions of $\bullet OH$ radicals [Taube and Bray, 1940; Allen, 1948]. The effect occurs due to radical ion transfer reactions of the type $\bullet OH + X^- = OH^- + X\bullet$. Thus, $X\bullet$ radicals can be left in the medium and are potential halogenating agents of organic compounds. These also can react directly with water: $X\bullet + H_2O = X^- + H^+ + \bullet OH$. (Similar reactions of $X\bullet$ radicals with hydrogen peroxide are indicated in Section 4.2.)

The thermodynamic data relating the reactions are reported as [Uri, 1952]:

	$\Delta H\bullet$ (kJ/mol)	$\Delta G\bullet$ (kJ/mol)
$X\bullet = F\bullet$	-88	-63 (exothermic)
$X\bullet = Cl\bullet$	$+41.8$	$+46$ (endothermic)
$X\bullet = Br\bullet$	$+96$	$+100$ (endothermic)
$X\bullet = I\bullet$	$+167$	$+163$ (endothermic)

These thermodynamic data, to which an activation energy must be associated, indicate that the probability of retroformation of $\bullet OH$ starting from $X\bullet$ is low. (In the case of the exothermic reaction of $F\bullet$, the activation energy in aqueous solution is estimated on the order of 20 to 40 kJ/mol.) Except for the reactions with hydrogen peroxide species commented on later, the most significant effects of radical ion transfer reactions are related to bicarbonate and carbonate ions often present at relatively high concentrations in drinking water.

Scavenging reactions reported are:

$$\bullet OH + CO_3^{2-} = OH^- + CO_3^{-\bullet}$$

and

$$\bullet OH + HCO_3^- = OH^- + HCO_3^{\bullet}$$

With carbonate ions, the effect is much more important than with bicarbonate ions [Hoigné and Bader, 1977]. The carbonate radical remains an oxidant by itself, but its capabilities in water treatment have not yet been explored thoroughly. For example, it is reported that when oxidations are promoted by hydroxyl radicals in the presence of bicarbonate–carbonate ions in the aqueous phase, the potential formation of bromate ion by oxidation of bromide-hypobromite is increased vs. bromate formation in the absence of bicarbonate–carbonate ions. As a preliminary design rule, one can state that carbonate ion is best absent in waters treated by methods based on •OH radicals (i.e., to operate at pH values lower than 8).

In aqueous solution, the $HO_2^•$ radical can dissociate into H^+ and $O_2^{-•}$. The pK_a value of $HO_2^•$ equals about 2 [Uri, 1952]. The molecular oxygen $O_2^{-•}$ monovalent ion radical in aqueous solution is a supposed intermediate in the H_2O_2/UV processes discussed later. The first electron affinity of oxygen (exothermic) is reported as 66 kJ/mol ($O_2 + e = O_2^- + 66$ kJ/mol). The mono-ion radical is solvated (solvation energy is proposed as 293 kJ/mol). Oxygen as a molecular divalent ion (O_2^{2-}) is hydrolyzed into HO_2^- and OH^- with an exothermic balance of +376.6 kJ/mol.

4.1.3 ANALYTICAL EVIDENCE OF •OH RADICALS
IN WATER TREATMENT

Bors et al. [1978] have considered the practical possibilities of evidence of the specific presence of •OH radicals under conditions comparable to those during the treatment of drinking water. Bleaching of *p*-nitrosodimethylaniline seems to be a possible method because the dye is not bleached by singlet oxygen [Kraljic and Moshnsi, 1978; Sharpatyi et al., 1978]. The solutions of the dye also are stable in the presence of hydrogen peroxide, but not with application of hydrogen peroxide + UV [Pettinger, 1992]. Ozone-free UV light does not bleach the dye within delays encountered in practice. Ozone, however, added or generated on-site, interferes.

p-Nitrosodimethylaniline reacts rapidly with hydroxyl radicals: $k_2 = 1.2 \times 10^{10}$ L/mol-sec [Baxendale and Wilson, 1957]. At pH = 9, the molar absorption coefficient in water, at 435 nm, has been reported as 84,400 L/mol·cm. It is recommended to measure the bleaching of a solution at the initial concentration of 4×10^{-4} mol/L, and to operate with water that is saturated in oxygen vs. air [Pettinger, 1992].

No precise protocol or standard method has yet been defined for the detection and determination of •OH radicals under conditions applicable to drinking water treatment processes. It must be remembered that the lifetime of hydroxyl radicals is in the range of nanoseconds and that the potential stationary concentration of radicals such as •OH in water is low (estimated 10^{-12} to 10^{-13} mol/L by Acero and von Gunten [1998]).

The absorbance of hydroxyl radicals in the UV-C range is about 500 to 600 L/mol·cm. Comparative values at 254 nm are 1000 L/mol·cm for $HO_2^•$; 2100 L/mol·cm for O_2^-; 150 L/mol·cm for $HO_3^•$. A general value for aliphatic peroxy radicals is in the range of 1200 to 1600 L/mol·cm. The case of hydrogen peroxide is mentioned later.

It can be concluded that the potential optical interference of such radicals under conditions of water treatment is negligible in the UV-C range. However, such radicals can be activated by absorbing UV-C light, and as such they cannot be neglected. An overview of literature on the degradation of chlorophenols is reported by Trapido et al. [1997].

4.1.4 Reactions of Hydroxyl Radicals with Organic Compounds in Aqueous Solution

Several mechanisms are operating in concomitant and competitive ways, as explored by Peyton [1990].

4.1.4.1 Recombination to Hydrogen Peroxide

The recombination to hydrogen peroxide reaction follows:

$$2 \; \bullet OH = H_2O_2$$

4.1.4.2 Hydrogen Abstraction

The hydrogen abstraction reaction is illustrated by:

$$\bullet OH + \cdots + RH_2 = ...,RH\bullet + H_2O$$

These first steps are followed by a reversible reaction with dissolved oxygen:

$$RH\bullet + O_2 = RHO_2^{\bullet}$$

Hydrogen abstraction seems to be the dominant pathway. *As a design rule, one can recommend the water to be saturated (even oversaturated) in dissolved oxygen concentration if submitted to •OH-based oxidations.*

The organic peroxyl radical $RHO_2^{\bullet}$ can further initiate thermally controlled oxidations.

- *Decomposition and hydrolysis*: $RHO_2^{\bullet} = RH^+ + (O_2^{-\bullet} + H_2O) = RH^+ + H_2O_2$
- *Homolysis*: $RHO_2^{\bullet} + ..., RH_2 = RHO_2H$ (i.e., hydroxyl, carbonyl, and carboxylic compounds) $+ RH\bullet$, thus initiating a chain mechanism; generation of polymer products also possibly occurring; the latter easily removed by classical processes like coagulation–flocculation–settling
- Deactivation by hydrolysis of $O_2^{-\bullet}$ into H_2O_2 thus maintaining another cyclic pathway

4.1.4.3 Electrophilic Addition

Direct addition to organic π-bond systems like carbon–carbon double bond systems, leads to organic radicals that are intermediates in dechlorination. An exhaustive review on chlorophenols is available [Trapido et al., 1997].

4.1.4.4 Electron Transfer Reactions

$$\bullet OH + RX = OH^- + RX^{+\bullet}$$

This reaction corresponds to the reduction of the •OH radical and seems to be important in the case of multiple halogen-substituted compounds.

4.2 COMBINATIONS OF HYDROGEN PEROXIDE AND ULTRAVIOLET LIGHT

4.2.1 GENERAL ASPECTS

Hydrogen peroxide can be present in natural waters at concentrations in the range of 0.01 to 10 μM (i.e., 0.34 $\mu g/L$ to 0.34 mg/L). This natural hydrogen peroxide can be decomposed by sunlight and can contribute to natural purification mechanisms. However, the reacting concentrations correspond to very low levels. Hydrogen peroxide is an allowed technical additive in drinking water, for example, at concentrations of 17 mg/L in Germany or 10 mg/L in Belgium. The European Commission of Normalization (CEN) is considering the adoption of a limit of 17 mg/L.

Advantages of hydrogen peroxide as a source of hydroxyl ions are:

- Wide commercial availability of the reagent
- High (almost infinite) miscibility with water
- Relatively simple storage conditions and dosing procedures
- High potential yield of production of hydroxyl radicals: two per molecule

Major specific disadvantages of the direct use of hydrogen peroxide in the •OH-based photochemical processes for water treatment are:

- Low absorbance in the classical UV range of wavelengths (*vide infra*)
- Potential disproportionation reactions to form hydroperoxyl radicals; the latter, (less or not active) putting a limit on the potentially useful hydrogen peroxide concentration that can be set in: $H_2O_2 + •OH = H_2O + HO_2^•$

The most commonly accepted mechanism of initial reaction of hydrogen peroxide to produce hydroxyl radicals on irradiation with UV light is the cleavage into two •OH radicals: $H_2O_2 + (h\nu) = 2 •OH$. The quantum yield is about unity in dilute solutions. According to the thermodynamics, this reaction phase is endothermic to the extent of about 230 kJ/mol. Activation energy remains necessary to maintain the internuclear distances during the photodissociation (Franck–Condon principle). The necessary initial energy input is in the range of 314 kJ/mol [Kornfeld, 1935].

At high concentrations (e.g., in the range of grams per liter), the direct UV photolysis of hydrogen peroxide is of zero order. In other words, under such conditions that exist in industrial applications, the photonic flux is the rate-determining step. At lower concentrations, up to concentrations of 10 mg/L of hydrogen peroxide, the dissociation reaction of hydrogen peroxide obeys first-order kinetics: $C(H_2O_2) = C_o(H_2O_2) \times e^{-kt}$. The k values can differ as a function of the UV lamp technology and reactor design. Typical values for k are, for example, 0.016/min for a low-pressure 8-W(e) lamp (without the 185 nm-line), and 0.033/min for a 15-W(e) lamp transmitting also the 185 nm-line. At similar electrical power input, the k value can be approximately doubled by Xenon-doped low-pressure mercury lamps also emitting a continuum around 200 to 220 nm [Pettinger, 1992].

Under this assumption the kinetic constants can be translated as [Guittonneau et al., 1990]:

$$k = (2.3 \times A \times \Phi \times L \times r \times I_0)/V$$

where

A = absorption coefficient (base 10)
Φ = quantum yield
L = layer thickness
r = (UV light) reflection coefficient of the reactor wall
I_0 = radiant intensity of the UV source
V = irradiated water volume

The quantum yield in the milligram per liter concentration range is reported as 0.97 to 1.05 [Baxendale and Wilson, 1957]. Therefore, measurement of the ratio of hydrogen peroxide photolysis under practical reactor conditions enables measuring the photon flux in a given lamp–reactor configuration as well as checking the constancy of operational conditions during a series of experiments [Guittonneau et al., 1990]. However, the quantum yield of hydrogen peroxide photolysis has been reported as dependent on temperature: $\Phi = 0.98$ at 20°C, and 0.76 near 0°C [Schumb and Satterfield, 1955]. The practical result is a necessary compromise between the drop in UV output as a function of the outside temperature of the lamp and the quantum yield.

Pettinger [1992] has repeated the experiments with a low-pressure lamp (Heraeus TNN 15) and normalized the first-order kinetic constant of decomposition of dilute aqueous solutions (10 ppm) of hydrogen peroxide vs. photon output of the lamp as a function of the temperature. A set of data is presented in Table 12.

At 253.7 nm, the absorption coefficient of H_2O_2 (base 10) equals 18.6 l/mol·cm, whereas for the (acid) dissociated form, HO_2^-, $A = 240$ l/mol·cm. Consequently, the acidity constant of hydrogen peroxide ($pK_a = 11.6$) can influence the yield of photochemical dissociation of dissolved H_2O_2 very significantly. In natural waters, however, high pH values (e.g., 12 and higher) do not occur. Because carbonate alkalinity is scavenging the •OH, a necessary compromise needs to be established on the basis of overall analytical data and the treatment objectives.

TABLE 12
First-Order Kinetic Constants for UV Decomposition Of H_2O_2 in Dilute Aqueous Solution vs. Photon Output

T (°C)	Relative Lamp Output	k Measured (min^{-1})	k/Relative Output
25.0	1.00	0.034	0.034
17.5	0.78	0.029	0.037
10.0	0.58	0.026	0.045
5.0	0.45	0.013	0.028

From Pettinger, 1992.

Additionally, disproportionation of hydrogen peroxide is known to occur at the pH of its pK_a value of 11.6, as follows:

$$H_2O_2 + HO_2^- = H_2O + O_2 + {}^{\bullet}OH$$

The absorption of hydrogen peroxide in the UV-C range is illustrated in Figure 92. Consequently, in presently available lamp technologies applicable to the scale of drinking water treatment, the doped lamps emitting the 200- to 220-nm continuum and medium-pressure lamps are the most performant in generating $^{\bullet}OH$ from aqueous hydrogen peroxide.

However, the secondary effect of nitrates needs to be considered in natural waters.

4.2.2 EFFECTS OF NITRATE ION CONCENTRATION

The absorption spectrum of the nitrate ion in aqueous solution is indicated in Figure 93. There is competition for absorption by nitrates, thus lowering the available photon dose and the yield of generation of radicals by the photodecomposition of hydrogen peroxide by UV light in the 200 to 230 nm range. This competition is higher for doped low-pressure Hg lamps also emitting in the 200 to 220 range than for the high-intensity, medium-pressure lamps.

By absorption of UV light, the nitrate ion is activated:

$$NO_3^- + h\nu = (NO_3^-)^*$$

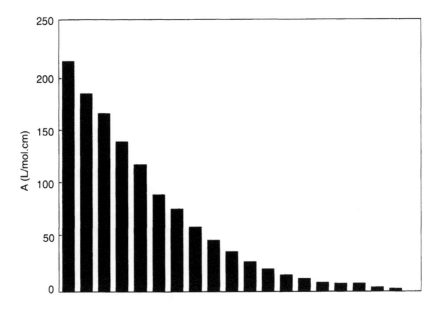

FIGURE 92 UV absorbance of H_2O_2 (abscissa is in 5-nm steps from 195 to 290 nm).

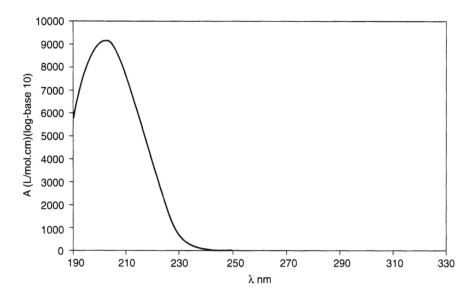

FIGURE 93 UV absorption spectrum of aqueous nitrate.

Subsequent formation of nitrite ion can result:

$$(NO_3^-)^* = NO_2^- + (O)$$

and:

$$(O) + NO_3^- = NO_2^- + O_2 \quad \text{[Bayliss and Bucat, 1975; Shuali et al., 1969]}$$

An alternative route to the formation of nitrite ion is accompanied by the formation of •OH according to Shuali et al. [1969]:

$$(NO_3^-)^* = NO_2^- + O^*$$

and:

$$O^* + H_2O = \text{•OH} + OH^-$$

The reactions occur over a wide range of pH (1.5 to 12.8).

The formation of nitrite ion by photolysis of nitrate ion, whether or not assisted by hydrogen peroxide, is illustrated in Figure 94 [Pettinger, 1992]. At low irradiation doses, the addition of hydrogen peroxide has a significant effect. At higher photon doses, however, the photochemical effects become predominant.

With lamps emitting the 185-nm wavelength and as well as in the 200- to 220-nm range, the formation of nitrite ions becomes important (as illustrated in a sample experiment shown in Figures 94 and 95). Formation of nitrite ion on exposure to indium-doped lamps is of another order of magnitude than with conventional low-pressure germicidal lamps.

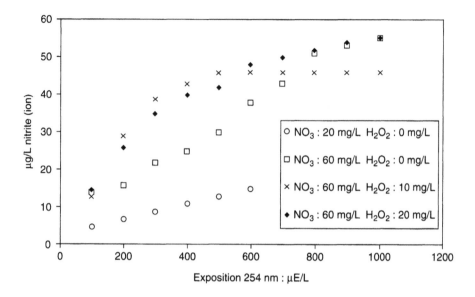

FIGURE 94 UV photolysis of nitrate ion. (Low-pressure Hg lamp with 185-nm wavelength filtered, classical germicidal lamps without increased emission yield in the 200 to 220 nm range.) (From Pettinger, K.H., thesis, Technical University of Munich, Germany, 1992.)

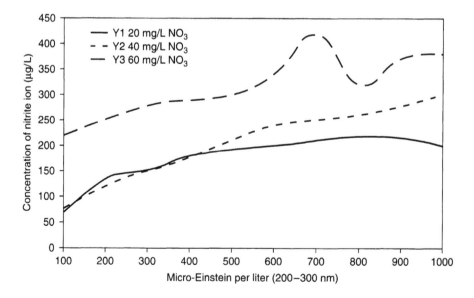

FIGURE 95 Increased formation of nitrite ion during 185- to 220-nm irradiation of nitrate.

4.2.3 Reported Data on Ultraviolet Synergistic Oxidation with Hydrogen Peroxide

Not many studies of full-scale data have been reported extensively yet on hydrogen peroxide-UV in drinking water, but the process is in significant development. Data from laboratory and pilot plant investigations on the removal of organic products from water have been reviewed by Legrini et al. [1993]. One has the following sources of information classified by type of compound oxidized:

Chlorinated (and brominated) low molecular-mass compounds—Glaze et al. [1987]; Guittonneau et al. [1988]; Masten and Butler [1986]; Sundstrom et al. [1986]; Sundstrom et al. [1989]; Symons et al. [1989]; Weir and Sundstrom [1989].

Phenols, chlorinated phenols and nitrophenols—Castrantas and Gibilisco [1990]; Köppke and von Hagel [1991]; Ku and Ho [1990]; Lipczynska-Kochany and Bolton [1992]; Sundstrom et al. [1989]; Yue and Legrini [1989, 1992]; and more related to waste treatment, Yost [1989].

Organochlorine pesticides—Bandemer and Thiemann [1986]; Bourgine and Chapman [1996, p. II.a]; Winner [1993]; Kruithof [1996]; Kruithof and Kamp [2000].

Monocyclic aromatic hydrocarbons (also substituted)—Barich and Zeff [1990]; Bernardin [1991]; Beyerle-Pfnür et al. [1989]; Cater et al. [1991]; Glaze and Kang [1990]; Guittonneau et al. [1988a,b, 1990]; Ho [1986]; Peterson et al. [1990]; Sundstrom et al. [1989]; Symons et al. [1989]; Symons et ai. [1990]; Bischof [1994]; Dussert et al. [1996, p. I.IV.a].

Miscellaneous compounds and special cases
  Carbon tetrachloride. Guittonneau et al. [1988]; Sundstrom et al. [1989]; Symons et al. [1989]
  Diethylmalonate. Peyton and Gee [1989]
  Dimethylhydrazine. Guitonneau et al. [1990]
  Dioxane. Cater et al. [1991]
  Freon TF. Yost [1989]
  Methanol. Zeff and Leitis [1988]; Barich and Zeff [1990]
  Herbicides. Peterson et al. [1990]; Pettinger [1992]

Often, the results of investigations remain system-dependent (a particular reactor and a given lamp are used). Although the geometric description of the systems generally is precise, it is not possible to accurately estimate the true *absorbed dose* in kilojoule per square meter, which is the necessary parameter for commercial scaling-up. Data are most often reported in terms of *exposure dose*. Recommendations on this point are formulated at the end of this chapter.

At laboratory temperatures of 20 to 22°C and with initial concentrations of pollutants in the range of 10^{-5} to 10^{-4} mol/L, a general observation of the results would be that an abatement of 60 to 100% requires 30 to 90 min. This also depends on the general matrix of the water (total organic carbon [TOC], pH, dissolved oxygen, etc.). In terms of energy consumption, in practice and also in pilot investigations one must reckon with 0.8 kWh for 80% removal [Bourgine and Chapman, 1996].

Major full-scale applications are planned, because the combination UV-H_2O_2 promotes less formation of bromate ion in bromide-containing water [Kruithof and Kamp, 2000].

4.3 SYNERGISM OF OZONE AND ULTRAVIOLET LIGHT IN WATER SANITATION

4.3.1 DECOMPOSITION OF OZONE BY ULTRAVIOLET IRRADIATION

Ozone strongly absorbs UV light with a maximum absorbance at 260 nm (i.e., at about the emission of low-pressure mercury lamps). This is the so-called Hartley-band illustrated in Figure 96 (maximum absorbance, 3000 L/mol·cm).

Again, as for the bactericidal effect, the potential efficiency of polychromatic UV sources for ozone-UV synergism can be calculated by a 5-nm histogram approach. For polychromatic sources, setting the maximum absorbance of ozone as 1 at 254 nm (A = 3000 L/mol·cm), one obtains the following relative potential efficiency parameters: $f(\lambda)$ = absorbance at λ segment: 3000; $I_o(\lambda)$ = the relative part of the intensity emitted in the 5-nm segment; and $f(\lambda) \times I_o(\lambda)$ = the potential efficient emission of the lamp in the segment to produce synergistic action with ozone (Table 13).

The direct efficiency of a broadband UV source emitting as illustrated in Figure 22 thus ranges as follows:

$$\sum f(\lambda) \times I_o(\lambda) = 27.8$$

This means that nearly 30% of the UV-C emitted intensity of such a lamp is potentially efficient for UV-ozone synergistic processes.

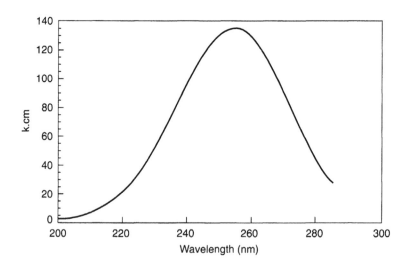

FIGURE 96 Absorption of UV light by ozone.

TABLE 13
Potential Efficiencies of UV Lamps to Produce Synergistic Efficiencies with Ozone

λ	$f(\lambda)$	$I_o\lambda$	$f(\lambda) \times I_o(\lambda)$	λ	$f(\lambda)$	$I_o\lambda$	$f(\lambda) \times I_o(\lambda)$
200	0.02	1	0.02	205	0.033	1.85	0.061
210	0.055	2.45	0.13	215	0.089	3.15	0.28
220	0.133	3.4	0.45	225	0.206	3.35	0.69
230	0.3	3.15	0.95	235	0.47	2.75	1.29
240	0.76	2.1	1.6	245	0.89	0.95	0.85
250	1	3.45	3.45	255	0.98	5.25	5.15
260	0.944	7.0	6.60	265	0.83	3.45	2.86
270	0.63	2.1	1.32	275	0.47	1.55	0.73
280	0.32	2.5	0.8	285	0.206	0.6	0.124
290	0.139	1.55	0.22	295	0.09	0.8	0.072
300	0.045	3.5	0.16	305	0.028	—	—
310	0.01	—	—	315	—	—	—

Numerous publications concern the decomposition of ozone by UV light, mainly in relation to atmospheric ozone. In summary, for water treatment purposes the photolysis of ozone is basically of first-order vs. the absorbed photon dose [Masschelein, 1977, Laforge et al., 1982]. The quantum yield (number of ozone molecules decomposed per absorbed photon) can vary between 4 and 16. It is strongly increased by the presence of water, either vapor or liquid (including droplets). The overall reaction scheme has been reported [Wayne, 1972; Lissi and Heicklen, 1973; Chameides and Walker, 1977]. Moreover, the quantum yield increases at increased ozone concentration in the gas phase. The whole is related to the energy diagram of oxygen (Figure 97), and the chain reaction mechanism is supported by two important exothermic reactions:

$$O(^1D) + O_3 = 2O_2 + 580 \text{ kJ/mol}$$

and:

$$O(^3P) + O_3 = 2O_2 + 400 \text{ kJ/mol}$$

4.3.2 PRACTICAL EVIDENCE

4.3.2.1 Mixed-Phase Systems

There is evidence that in mixed-phase systems (i.e., gaseous ozone is bubbled through water contained in or flowing through a UV reactor), the reactions occur in the boundary layer of the gas–liquid interface instead of the bulk of the solution [Denis et al., 1992]. This can be a way of direct action of the $O(^1D)$ radical on the molecules that are oxidized.

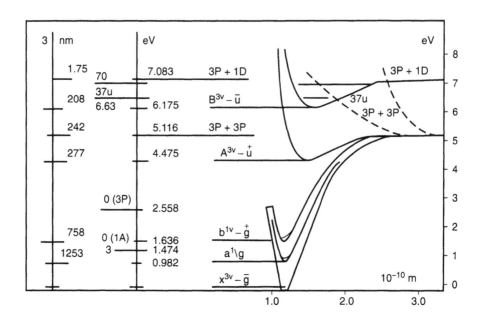

FIGURE 97 Energy diagram of oxygen.

4.3.2.2 Homogeneous-Phase Systems

Another hypothesis for the decomposition of dissolved or aqueous ozone (i.e., in homogeneous phase by UV) is the intermediate formation of hydrogen peroxide [Peyton and Glaze, 1986, 1988]: $O_3 + h\nu = H_2O_2 + O_2$. From this reaction on, hydroxyl radicals can be formed, as considered in the discussion of hydrogen peroxide-UV systems.

The combined ozone-UV processes have been used for decolorization of pulp bleaching waters in the paper industry [Prat et al., 1990]. At present, the technique is widely used to treat industrial effluents and landfill leachate water [Leitzke, 1993]. Application to drinking water treatment for the removal of toxic or hindering compounds can be expected in the future, but a cost-efficiency evaluation is necessary in each individual case. Sierka et al. [1985] intensively studied the removal of humic acid-TOC with medium-pressure Hg lamps—99% removal could be achieved after 50 min of irradiation. A wide variety of compounds capable of being oxidized by synergistic UV ozonation has been reported by Legrini et al. [1993].

Also an exhaustive list contains primarily:

- Chloroaliphatic compounds [Francis, 1986; Himebaugh and Zeff, 1991]
- Chlorinated aromatics [Fletcher, 1987]
- Phenolic compounds [Gurol and Vastistas, 1987; Trapido et al., 1997]
- Substituted aromatic compounds [Xu et al., 1989]
- 2,4-Dichlorophenoxyacetic acid or 2,4-D, [Prado et al., 1994]
- Alcohols, carboxylic acids, and aldehydes [Takahashi, 1990]
- Pesticides [Yue and Legrini, 1989]

- Cyanazine herbicide [Benitez et al., 1994]
- Detergents, dyestuffs, and chlorinated benzenes [Shi et al., 1986]
- Glycols [Francis, 1986]

4.3.3 Costs

So far, for financial reasons, the full-scale application of combined UV-ozone treatment processes on large flows such as in drinking water treatment remains limited. Major developments can be expected, however. The method is currently applied for the sanitation of heavily charged industrial effluents through lixiviation.

4.3.4 Technological Generation of Ozone by Ultraviolet Irradiation of Oxygen (or Air)

The potential for photochemical generation of ozone for water treatment has been claimed, and the scope and limitations of the method using conventional UV lamps has been reviewed extensively [Dohan and Masschelein, 1987]. The first information relating to the formation of ozone from oxygen exposed to UV light in the region of 140 to 190 nm was reported in Lenard [1900] and fully assessed by Goldstein [1903]. The practical yields that were obtained in these early experiments were in the range of 50 to 300 mg/m^3 ozone in oxygen with an energy balance of 0.2 to 0.3 g/kWh, or 3 kWh/g O_3. Methods are claimed by McGregor [1986].

Basically, with conventional or advanced Hg lamp technologies, the generation of ozone relies on 185-nm resonance emission. However, the absorbance of oxygen is weak (Figure 98).

A preliminary conclusion can be formulated as follows:

- The absorbance (base 10) of oxygen at 185 nm is about 0.1 cm^{-1} atm^{-1}; in other words, it is much lower than the Hartley absorbance of ozone in the 260-nm region (134 cm^{-1} atm^{-1}), so that the photostationary balance of ozone formation vs. ozone decomposition with traditional Hg lamps is unfavorable.
- Optical filters, as existing (see Section 2.7.6, Chapter 2) cut off emissions at lower UV wavelengths, thus balancing the photostationary equilibrium in less favor of ozone formation vs. ozone photolysis.
- If by doping the Penning gas a hypsochromic shift of the emission could be achieved, an increase in the absorbance by oxygen at a given wavelength by a factor of 10 to 100 could result (see Figure 98); this could reverse the photostationary balance.
- Developing continuously emitting UV lamps on the basis of xenon emission gas is a challenge for future improvement [Fassler and Mehl, 1971].
- Unpublished observations [Bossuroy and Masschelein, 1987 to 1992] show, however, that aging of the quartz of the lamp or the enclosures of mercury lamps emitting at the 185-nm wavelength (so-called *ozone positive* lamps) occurs very fast, resulting in loss of transmission at these wavelengths by at least 50% after 500 to 700 h. This observation, to be confirmed, can be of significant importance in the solarization of the quartz and the enclosure

material of UV lamps. Ozone-free lamps are obtained by incorporating ozone inhibitors into the quartz (e.g., titanium dioxide).

- Xenon excimer lamps are proposed for ozone generation in an oxygen flow in the controlled vacuum UV (VUV) range. The process [Hashem et al., 1996] initiating oxygen–ozone-related oxidants at wavelengths lower than 200 nm is promising. Aging of hardware is not yet thoroughly established.

4.4 ULTRAVIOLET CATALYTIC PROCESSES

A tentative development in oxidation applied for the removal of resistant contaminants is the photocatalytic process using radicals produced by absorption of light on semiconductors. Carey et al. [1976] were the first to report on photocatalytic degradation of biphenyl and chlorinated biphenyls in the presence of titanium dioxide.

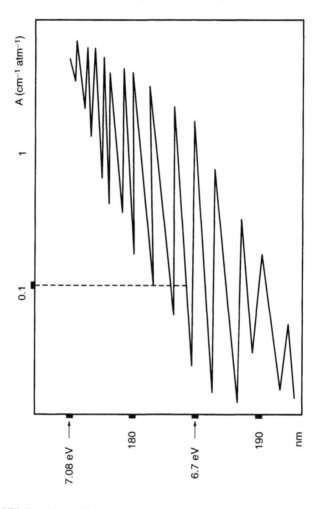

FIGURE 98 UV absorbance of gaseous oxygen.

Titanium dioxide, whether or not in association with other metal oxides, has since been prospected intensively for the removal of hindering compounds from leachates of disposed hazardous wastes. Literature on the subject is reviewed by Legrini et al. [1993] and in proceedings of seminars [Al Ekabi, 1994 et seq.]. As far as drinking water treatment is concerned, no significant applications exist so far. The association of other metal oxides with TiO_2 (Ce, Pt, Mn, etc.) may require preliminary toxicological evaluation for this application. Association with copper ions is another feasible option [Paillard, 1996].

The basic principle of the process is the production of a semiconductor having an electron hole:

$$TiO_2(+h\nu) \rightarrow TiO_2(h^+) + TiO_2(e^-)$$

This reaction can be initiated by all wavelengths of the conventional UV ranges, A, B and C. Consequently, all light emitted by medium-pressure Hg lamps is potentially applicable.

The quantum yield of the initiating reaction is very low, either less or equal to 0.05 according to Mathews [1989], Pellizetti [1985], and Bolton [1991]. The reaction can be obtained with some limits on exposure to sunlight [Takahashi et al., 1990]. The oxidations at the electron-depleted entity only occur at short distances, that is, to molecules *adsorbed on the catalyst* (e.g., water or adsorbed organics):

$$TiO_2(h^+) + H_2O \text{ (adsorbed)} \rightarrow TiO_2 + H^+ + \bullet OH \text{ (adsorbed)}$$

Direct reactions of adsorbed organic compounds are possible as well:

$$TiO_2(h^+) + RY \text{ (adsorbed)} \rightarrow TiO_2 + H^+ + RY\bullet \text{ (adsorbed)}$$

Molecular oxygen must be present to act as the electron acceptor:

$$TiO_2(e^-) + O_2 \rightarrow TiO_2 + O_2^{-\bullet}$$

Complementary addition of hydrogen peroxide can increase considerably the formation of $\bullet OH$ radicals, supposedly through following route:

$$TiO_2(e^-) + H_2O_2 \rightarrow TiO_2 + OH^- + OH\bullet$$

Determining parameters are pH, oxygen concentration, and TOC concentration. The process is controlled by adsorption and reactions at short distances on the catalyst surface. Therefore, the major problems to be solved are related to the reversibility of the adsorption, the construction of adequate catalyst surfaces, the removal of the catalyst from the treated water, the materials of construction of the equipment, and the neccessary fundamental knowledge on the kinetics of the processes occurring [Legrini et al., 1993], [Okatomo et al., 1985a,b]. Up to now, the titanium dioxide processes have been focused on the treatment of industrial effluents and leachates, but developments for drinking water treatment are likely to occur in the future.

4.5 TENTATIVE DESIGN RULES FOR ULTRAVIOLET SYNERGISTIC OXIDATION PROCESSES

As explained in Section 4.1.1, the data reported—even very precise—are often too system-dependent to enable direct formulation of general guidelines. An attempt is made to draw the attention to some essential aspects that necessarily precede design work of UV-assisted oxidation processes, as follows:

1. *Flow of water to be treated, treatment scheme already existing*—Minimum, maximum, and average guidelines are expected for the future.
2. *Complete recorded and quantified UV spectrum of the water to be treated*—These spectra (with variations in time, if any) in the full UV range from 200 to 400 nm are a dominant parameter in the choice of the type of lamp technology and possibly of the oxidant to be used in conjuction with UV.
3. *A record of the total organic carbon (TOC), with variations in time*— More precise information on the specific type of contaminants, as present and whose removal is the objective, is of utmost important.
4. *Ionic balances of the water composition and variations with special emphasis on*
   a. Total alkalinity, carbonate alkalinity and pH. These parameters can determine the necessity of installing an acid pretreatment.
   b. Dissolved oxygen concentration. UV-assisted oxidation processes are favored by high dissolved oxygen concentrations; the concentrations as existing can determine the need for a preliminary aeration or oxygenation.
   c. Nitrate ion content of the water, and variations. This parameter directly influences the choice of lamp technology and also can determine the choice of UV-H_2O_2 or UV-O_3 technologies.
   d. Temperature range of the water. This parameter influences the choice of the lamp technology to apply and possibly the lamp enclosure.
   e. Turbidity range of the water. This parameter influences the lamp cleaning procedures of the system.
5. *Total plate counts (TPC) and enterobacteria number*—These numbers can determine the combination of disinfection–oxidation.
6. *Preliminary investigations*—If feasible within the economics of the local conditions, it is recommended to make at least a preliminary investigation in the laboratory, preferably a pilot study on the water to be treated. To be useful for design, these evaluations should report the information specified in 2 to 5. Moreover, besides a full description of the method, these evaluations should describe the irradiation dose applied (joule per square meter). Concepts of experimental reactor designs and evaluation of the dose are described in Chapter 2.
7. *Checklist of design of UV-synergistic oxidation processes in water sanitation*
   a. Reliability of preliminary information and definition of the objectives to achieve (see preceding 1 to 6)
   b. Basic justification of the system proposed vs. alternatives

c. Comments on complementary installation and operation conditions (incorporating into an existing or future treatment scheme pH-adjustment, aeration–oxygenation, mixing conditions, etc.)
d. Degree of automation and remote control required
e. Level of safety and redundancy required
f. Priority of objectives (either with or without combined disinfection and photochemical oxidation)
g. Justification vs. alternative oxidation methods
h. Standards and codes of practice to apply (electrical, ambient, and environmental safety aspects)
i. Economic evaluation of the project
j. Maintenance and costs of spare parts
k. Level of guarantee applicable
l. Reference to existing installations (location, publications, etc.)
m. Flexibility for future extensions and developments

5 Use of Ultraviolet Light for Sanitation of Wastewater

Ultraviolet (UV) light is a valuable alternative for disinfection of treated wastewater, because it forms no or very low levels of disinfection by-products. Among the negatives of the method that have been considered is the potential reactivation of organisms after exposure, whether or not in relation to shielding of organisms by suspended solids. At present, no general rules exist for the necessary (high) UV doses that could promote formation of by-products. Pilot investigations are advisable for each particular case. The potential toxicity of the treated effluent must be evaluated.

In contrast with drinking water treatment, a wastewater method is better established in the United States than in Europe. A survey made for the U.S. EPA [1986] found more than 600 utilities using UV for disinfection of secondary effluent, with the period of experience more than 20 years [Martin, 1994]. This development still is in progress, with the growing importance of the issue of disinfection by-products, but 1200 stations were mentioned to be in operation in the United States and Canada in 1995 [Blatchley and Xie, 1995]. No clear report is available on the number of European applications in wastewater treatment.

5.1 REGULATIONS AND GUIDELINES FOR DISINFECTION OF TREATED WASTEWATER

Concerning wastewater reuse for the purpose of irrigation of crops, the World Health Organization (WHO) recommends a maximum limit of 100 total coliforms per 100 mL, in 80% of the samples collected at regular intervals.

The Council Directive of the European Union concerning urban wastewater treatment (91/271/European Economic Community [EEC]) (O.J. 25-05-1991) does not require specific disinfection of treated wastewater as it is discharged into the environment. The member stated or the local authorities can lay down specific requirements as a function of reuse of treated water (recreation, shellfish culture, irrigation of crops...).

The directive of the (European) council of December 8, 1975 lays down the following bacteriological criteria for swimming water. They can be a good starting point to evaluate disinfected wastewater:

- *Total coliforms*—Guide number less than 500 per 100 mL for 80% of the samples at a given site, and imperatively less than 10,000 per 100 mL for 95% of the determinations at a given sampling site
- *Fecal coliforms*—Guide number less than 100 per 100 mL for 80% of the determinations and imperative criterion of less than 2000/100 mL for 95% of the determinations
- *Fecal streptococci*—At least 90% of the samples in compliance with the guide number of less than 100 per 100 mL

The directive is the basis of national regulations.

In France, general conditions of discharge and reuse of treated wastewater are defined by the Décret 94-469 of June 3, 1994. For specific reuse, permits remain case-dependent. For example, in the sea bathing station of Deauville, France, local criteria applicable (using chlorine dioxide) for discharge of secondary effluent during the summer period is less than 2000 total coliforms per 100 mL, with the effluent discharged at 2 km into the sea [Masschelein, CEFIC, 1996].

Another example involves Dieppe, France: Requirements have been set (for 95% of minimum 24 analyses) at total coliforms <10,000 per 100 mL, fecal coliforms <10,000 per 100 mL, *Streptococcus faecalis* <1000 per 100 mL [Baron et al., 1999]. ATV [1993], for example, also gives some general national recommendations.

In South Africa, the standards applicable to treated sewage specify the absence of fecal coliforms per 100 mL sample (see South African General and Special Standards [1984]).

In the United States, requirements are formulated by the U.S. EPA Design Manual on Municipal Wastewater Disinfection [Haas et al., 1986]. Again, the individual states can set specific requirements. Typical examples are cited next.

California regulations according to Title 22, Division 4, Chapter 3 of the California Code of Regulations follow:

- If used for spray irrigation of crops the median is less than 2.2 total coliforms per 100 mL (maximum allowed exception: less than 23 per 100 mL once a month) [Braunstein et al., 1994].
- The Contra Costa Sanitary District requires less than 240 total coliform bacteria per 100 mL [Heath, 1999]. At other locations, the local permit for total coliforms most probable network (MPN) is 23 per 100 mL as a monthly median with an allowable daily maximum of 500 per 100 mL.
- Gold Bar Wastewater Treatment of secondary effluent permits less than 200 total coliforms per 100 mL; tertiary effluent, less than 2.2 (MPN) total coliforms per 100 mL.
- Mt. View Sanitary District allows a 5-d median limit of 23 (MPN) total coliforms per 100 mL with a wet weather maximum of 230 per 100 mL.

In Florida, State Rule 62-600.400 of the Florida Administrative Code permits an annual average of less than 200 fecal coliforms per 100 mL, and no single sample containing more than 800 per 100 mL. In Massachusetts, the standard for average fecal coliforms for swimming water is less than 200 per 100 mL; in open shellfish areas, median total less than 70 per 100 mL (10% not exceeding 230 per 100 mL).

In Israel, the bacteriological criteria for reuse of treated wastewater in agriculture (and related applications) have been reviewed extensively [Narkis et al., 1987]. On the basis of 80% of the collected samples and per 100 mL, the limits for total coliforms for irrigation are set as:

- Less than 250 for vegetables to be cooked, fruits, football fields, golf courses
- Less than 12 for unrestricted irrigation of crops
- Less than three for irrigation of public parks and lawn areas (in 50% of the samples)

(In this context, the EEC Directive 75/440 on quality of surface water sources intended to be treated to obtain drinking water, recommends the following for the lowest quality allowable: total coliforms 500,000 per liter; fecal coliforms 200,000 per liter; fecal streptococci 100,000 per liter. The AWWA recommendations [AWWA, 1968] are less tolerant: total coliforms <200,000 per liter, fecal coliforms <100,000 per liter.

Most requirements in force concern enterobacteria (mostly coliforms). Counting of fecal coliforms is sometimes considered as an extended test. Some alternative tests have been considered, however, without general limits of tolerance. Proposed test organisms are bacteriophage f-2 (or MS-2) [Braunstein, 1994], and poliovirus seeded into the effluent [Tree, 1997]. *Clostridium perfringens* spores were also taken as an indicator for more resistent organisms (e.g., viruses) [Bission and Cabelli, 1980].

The estimated fecal coliform concentrations per 100 mL of undisinfected effluents are as follows (according to U.S. EPA): primary effluent, 10^6 to 10^7; secondary effluent, 10^4 to 10^5; and tertiary effluent, 10^3 to 10^5.

Figure 99 is a photo of UV disinfection of wastewater at the wastewater treatment plant at Gwinnett County, Georgia.

5.2 GENERAL CHARACTERISTICS OF EFFLUENTS IN RELATION TO DISINFECTION BY ULTRAVIOLET LIGHT

Dominant parameters to be considered are UV transmittance (UVT) and total suspended solids (TSS). As for the UVT, the wavelength of 254 nm is generally considered in the published articles. (This holds for the low-pressure Hg lamps; appropriate correction factors apply in the use of other lamp technologies [e.g., by the 5-nm histogram approach discussed earlier for drinking water disinfection].) The percentage of transmission is expressed for a layer thickness of 1 cm, and in terms of Beer–Lambert law on Log base 10 scale (sometimes not explicitly defined).

The unfiltered transmittance of a secondary-treated effluent is reported [Lodge et al., 1994] to be in the range of 35 to 82% (average 60%). From other literature sources, a range from 58 to 89% is observed and an average of 72% is probably suited in design [Appleton et al., 1994]. Acceptance of a value of 69.5% (to be

FIGURE 99 Disinfection of wastewater at Gwinnett County, Georgia. Total flow = 1580 m³/h, T_{10} = 74%. Each of four reactors is equipped with 16 medium-pressure lamps.

confirmed on-site) means an extinction value of $E = 0.4$ cm^{-1} and an absorbance value of $A = 0.15$ cm^{-1}, which are generally the first approximation values considered.

Suspended particles can exert several effects on the application of UV:

- Increase of optical pathway by scattering [Masschelein et al., 1989]
- Shielding of microorganisms
- Occlusion of microorganisms into the suspended material

The turbidity of unfiltered urban wastewater usually ranges between 1.5 and 6 units nephelometric turbidity units (NTU), but sudden surges can occur during run-off periods. The values for filtered wastewater range between 1 and 2 units (NTU). For wastewater, no general correlation exists between turbidity and suspended solids [Rudolph et al., 1994].

In domestic wastewater, the instant concentration of suspended solids usually is in the range of 600 to 900 g/m³. After 1-h static settling, it is in the range of 400 to 600 g/m³ (again, surges can occur, e.g., in the Brussels area up to 1000 g/m³). Globally, in urban sewage one can estimate the total suspended solids by 600 g/m³ on an average basis. About two-thirds are settleable (1 h). Of the remaining (average) 200 g/m³, about two-thirds are organic and one-third is mineral suspended solids.

Suspended solids in untreated wastewater usually present a bimodel distribution (Figure 100) with a maximum for particle diameters of submicron size and another maximum at 30 to 40 μm. With membrane filtration (1-μm pore size), the first maximum remains practically unchanged, whereas the second is lowered, however,

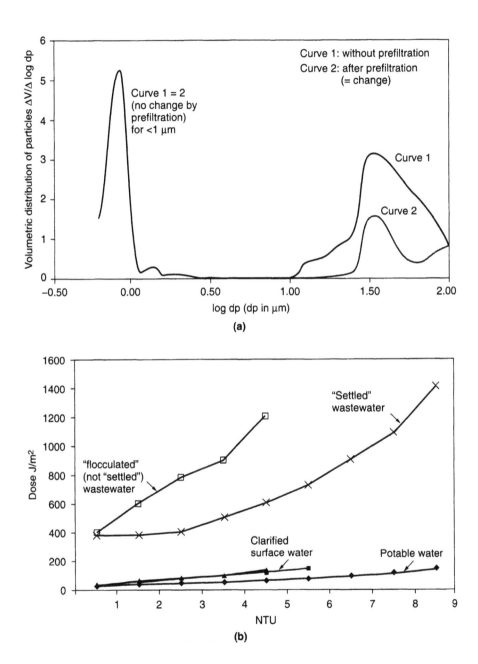

FIGURE 100 (a) Particle size distribution in secondary effluents; (b) effect of turbidity on the required dose (1, without prefiltration; 2, after prefiltration).

not completely removed. With intense mechanical mixing (estimated velocity gradient, $G = \geq 1000$ sec^{-1}) or ultrasonication, the large particle size material ($1.5 - 1.6$ μm) of the initial bimodel distribution can be partially destroyed as well as agglomerated to develop a trimodal distribution with secondary maxima at d_b at 0.1 to 0.2, 0.8 to 0.9, and 1.4 to 1.7 μm. This point might be important in laboratory experiments. More literature on particle-associated coliforms has been reported extensively by Parker and Darby [1994].

Overall, according to the data of Geesey and Costerson [1984], 76% of the bacteria are free-swimming and 24% are particle-associated. It is also reported that fecal bacteria adsorbed on sediments [Roper and Marshall, 1978], are more resistant to aggressions than free-swimming bacteria (e.g., irradiation by sunlight). Particle-associated bacteria are mostly found on suspended solids of particle diameter size larger than 10 μm [Ridgway and Olson, 1981, 1982].

It is not easy to establish a clear difference between adsorbed microorganisms, shielded microorganisms, and embedded microorganisms. A recommended procedure as published by Parker and Darby [1994] follows:

- Blend the sample (either wastewater or made-up sample) with an amphoteric detergent (e.g., Zwittergent) to make the concentration 10^{-6} M.
- Add a complexing agent (e.g., ethylenediaminetetraacetic acid [EDTA]) to make the sample at 3 to 12×10^{-3} M.
- Make it 0.01% (wt) in *tris*-peptone buffer.
- Adjust to pH 7 by phosphate buffering.
- Stir, operating at 19,000 r/min (about 320 r/sec) for 5 to 17 min. (The description is too vague to define a strict velocity gradient for the mixing conditions. From general methods of evaluation [Masschelein, 1991, 1996], the velocity gradient must have been higher than 5000 sec^{-1}.) Under such conditions of mechanical mixing, an apparent increase in total coliform counts by a factor of 4.0 to 7.7 could be observed. This means that the apparent direct numeration in the raw water can be a considerable underestimation of the total number if no vigorous agitation is applied on sampling.

Under static conditions (i.e., without mechanical mixing but by dosing the blending solutions only in static conditions) no significant apparent increase in counts of total coliforms was observed.

5.3 AFTERGROWTH AND PHOTOREPAIR AFTER EXPOSURE TO ULTRAVIOLET DISINFECTION OF WASTEWATER

It is difficult to distinguish between aftergrowth and photorepair in treated wastewater. In the first case, residual undamaged bacteria develop in the wastewater, which remains a nutrient medium. In the second case the schematic is as described in Chapter 3.

Note: In experimental work using artificial irradiation to promote photorepair, the mechanism is most often termed *photoreactivation*.

The generally proposed hypothesis is that a photoreactivating enzyme forms a complex with the pyrimidine dimer, the latter complex subject to photolysis by UV-A photons and restoring the original monomer as reported [Lindenauer and Darby, 1994; Harm, 1980; Jagger, 1967]. Visible light from UV up to 490 nm is also reported as able to promote photorepair. In other interpretations, enzymatic repair is considered to be possible in the dark [Whitby et al., 1984].

Many organisms have been found able to photorepair UV-damaged DNA, including total and fecal coliforms, *Streptococcus feacalis, Streptomyces, Saccharomyces, Aerobacter, Micrococcus, Erwinia, Proteus, Penicillium,* and *Neurospora*. On the other hand, some organisms have been reported not to be subject to photorepair: *Pseudomonas aeruginosa, Clostridium perfringens, Haemophilus influenzae, Diplicoccus pneumoniae, Bacillus subtilis,* and *Micrococcus radiodurans*. Literature is extensively reviewed by Lindenauer and Darby [1994].

There are several ways to quantify the photorepair:

N = concentration of organisms surviving UV disinfection

N_o = concentration of organisms prior to UV disinfection

N_{pr} = concentration of organisms after photorepair

Kelner [1951] defines the degree of photorepair by $(N_{pr} - N)/(N_o - N)$. To evaluate the possible photorepair in wastewater treated by UV-C, a log-increase *approximation* is more often used:

$$\log(N_{pr}/N_o) - \log(N/N_o) = \text{Log}[(N_{pr}/N_o)/(N/N_o)] = \log(N_{pr}/N)$$

According to literature, photoreactivation (in the log expression) could range between 1 and 3.4. However, photorepair and photoreactivation are related to the initial UV-C disinfecting dose. If the disinfecting UV dose is not sufficiently high, repair is greater. In the log approximation, no clear relation between the initial UV disinfecting dose and the yield of repair is obvious. By analyzing the data and expressing them in terms of degree of photorepair, however, a clear correlation is obtained (Figure 101).

No reported standardized testing procedures exist for evaluating photorepair or photoreactivating in water treatment. The use of white-light sources has been described by Lindenauer and Darby [1994] (e.g., a 40-W Vitalight source was used [Durolight Corp.]), placed at 75 cm over a layer of 1 cm of wastewater. The exposure was estimated at the exposure of 1 h sunlight at 12 noon (in the Californian sky).

The present conclusions on photorepair include:

- In wastewater disinfection by UV, a more careful analysis indicates that the photorepair is related to the UV exposure dose for disinfection, although in some publications, no relation between disinfection exposure dose and potential photorepair has been claimed.
- In practical conditions, the apparent regrowth as counted could also result from embedded organisms in the suspended solids.
- As indicated, some organisms are more subject to repair than others.

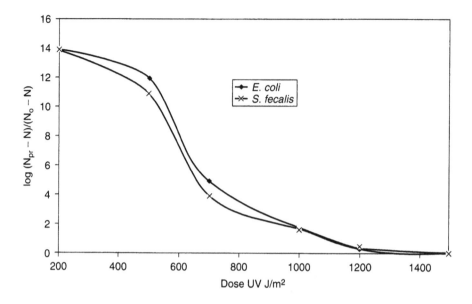

FIGURE 101 Photorepair after 1 h exposure to sunlight 40 W (total) on 1-cm thickness (based on data recalculated from measurements of Harris et al. [1987]).

- Indications exist that germs in nitrified effluents are more able to photo-repair than germs in unnitrified effluents.
- Practically all investigations concern the effects of low-pressure Hg lamps on DNA. In case of more general cellular destruction, probably occurring with high-intensity, medium-pressure Hg lamps, repair is less probable and not merely confined to DNA alone (see also Chapter 3, Section 3.2.3).

5.4 APPLIED ULTRAVIOLET DOSES IN WASTEWATER DISINFECTION

Most reported experiences thus far concern low-pressure Hg lamps, but the application of multiwave medium-pressure lamps is on the move. Because wastewaters are not constant in characteristics, the general recommendation is to make a sufficient pilot plant evaluation. Generally proposed exposure doses are 1000 to 1700 J/m^2 for general secondary effluent and 3000 J/m^2 for a nitrified effluent [Heath, 1999; Braunstein, 1994; Te Kippe et al., 1994]. The precise exposure doses are often not reported in a way that could allow generalizations. Some empiricism (or commercially restricted communication of know-how) remains in published information. The permanent control of the doses still relies on relative indications of a detector (generally a photocell), which also needs periodic calibration.

Besides the general quality of the wastewater, the necessary dose depends on the required level of organisms authorized by regulations, and the type of steering organism selected; and also in all this context, it must be remembered that the linear decay law usually applies only at high initial concentration of germs in the effluent. A tail-off occurs in the decay, as illustrated in Figure 102(a) and (b).

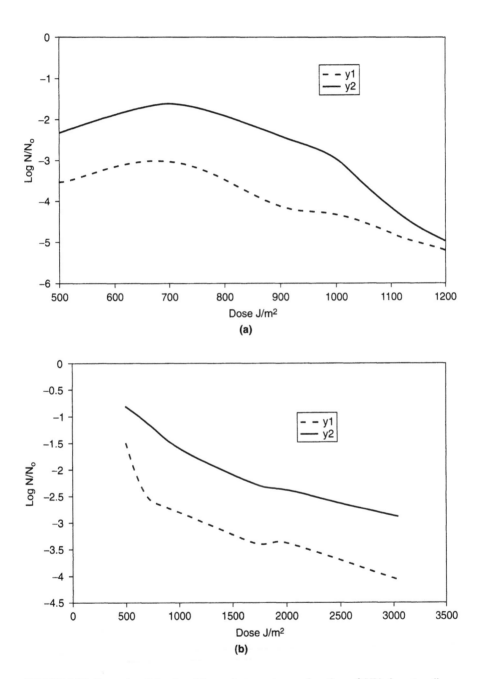

FIGURE 102 Example of fecal coliform abatement as a function of UV dose (medium-pressure Hg lamp). y1 = UV followed by solar illumination; y2 = solar illumination followed by UV. (a) Upper curves: nonnitrified, nonfiltered secondary effluent; (b) lower curves: nitrified, nonfiltered secondary effluent.

An empirical design model has been proposed as follows by Appleton et al. [1994]:

$$N = (f)D^n$$

where

N = bacterial concentration
D = active UV dose
f and n = empirical coefficients

The dose is estimated to be the average germicidal UV intensity (I) × irradiation time. The water quality factor f is approached by $f = A \times (TSS)^a \times (UVT)^b$, where A, a, and b are again empirical coefficients.

The whole is combined in an empirical model in which e is the random error of the model:

$$\log N = \log A + a \log(TSS) + b \log(UVT) + n \log I + n \log t + (e)$$

As for the average germicidal UV intensity again, an empirical binomial approach is considered:

$$I = -3.7978 + 0.36927\ (UVT) - 0.0072942\ (UVT)^2 + 0.0000631\ (UVT)^3$$

in which UVT is the UV transmittance in percentage of the unfiltered effluent. This approach was obtained for the Discovery Bay WWTP, California. It is not entirely established yet to what extent it can be of general value. However, the whole approach, based on the requirements for admissible limits for N and historical knowledge of TSS and UVT, ends in the choice of values for N and t.

The general structure of the method gives satisfactory results as reported; however, the essential parameters of the model can remain case-dependent. For the rest of design remaining determinants include hydraulic conditions, quality standards to be met and lamp technologies, intensity vs. irradiation time [Zukovs et al., 1986], maintenance, and performance control.

Numerous publications report on the installation of the lamps in the longitudinal mode (i.e., horizontal length in the same direction of the water flow [see Baron et al., 1999]), in the vertical mode (i.e., lamps up-down in the water flow [see Chu-Fei, H. Ho et al., 1994]). For low-pressure Hg lamps, these options appear not to be determinant in terms of efficiency. The choice parameters are related to both preexisting hardware to be retrofitted and general facilities for maintenance.

The Morrill index in comparable arrangements is about the same: between 1.15 and 1.35 in existing reactors [Blatchley et al., 1994]. The aspect ratio is usually higher in the horizontal lamp arrangement than in the vertical one. The aspect ratio A_R is defined by the following relation [Soroushian et al., 1994]:

$$A_R = X/L = X/4R_H = (X \times A_W)/4V_v$$

where

X = length of the reactor-contact basin into the direction of water flow
L = cross section of the UV lamps module perpendicular to the water flow ($L = 4R_H$)

R_H = hydraulic radius = (V_v/A_w)
V_v = *net* wetted volume that contains the lamps
A_w = *total* wetted horizontal surface of the module that contains the lamps

In existing plants with low-pressure Hg lamp technologies, the aspect ratio generally is between 15 and 40. The higher the value of A_R, the closer plug-flow conditions are approached. It is important to consider this parameter in designing pilot experiments, particularly for retrofitting plants in which existing basins will be used to install UV units for disinfection.

For high UV emission intensity technologies such as medium-pressure lamps, installation of the lamps in the vertical or traverse mode orthogonal to the water flow is preferred, both for facility of maintenance and for compact hardware. Mixing conditions and intensity distribution patterns are illustrated in Chapter 3 (Figures 80, 81, 82).

5.5 CHOICE OF LAMP TECHNOLOGY IN WASTEWATER DISINFECTION

In wastewater treatment, most present and existing applications are based on low-pressure lamp technologies. These are a result of historical factors related to technologies available at the time. From investigations [Kwan, 1994], medium-pressure high emission intensity systems can be more economical than the more conventional low-pressure lamp systems in both capital investment and lifetime costs (see also Soroushian [1994]). The number of plants making use of medium-pressure lamps is increasing rapidly. Until now, the use of excimer lamps and pulsed Xenon lamps in the field of wastewater disinfection remains experimental (e.g., for disinfection of agricultural wastewater [Hunter et al., 1998]).

A rule of thumb is to install 40 to 60 low-pressure lamps per 150 m^3/h of wastewater with an electrical power requirement of 65 to 80 W each. The electrical cost thus amounts to about 17 to 32 W/m^3. In some advanced installations, it can go up to nearly one lamp of 65 W(e)/m^3/h [Baron et al., 1999]. Low-pressure mercury lamps used in this application usually have a length between 1.2 and 1.5 m.

Note: Low-pressure mercury lamps operate only on an all-or-nothing on–off basis vs. the nominal emission capacity.

As described before, the output of medium-pressure lamps currently can be monitored between 60 and 100% of nominal emission capacity. This makes them attractive for treatment of variable water flows.

As for drinking water treatment, the lamps are installed in a quartz enclosure, which usually is mechanically cleaned with a to-and-fro wiper operated continuously or in an automated mode actioned as a function of a drop in light intensity as continuously measured.

Note: In the case of wastewater it is also necessary to clean the photocells and occasionally to recalibrate the system.

This mechanical cleaning procedure is more complex in the case of low-pressure Hg lamps, so that chemicals usually are required. A general cleaning procedure is to remove the lamps + enclosures per entire modules of several lamps and to dip them into an acid solution. The generally recommended solution is composed of phosphoric acid at 10% by weight. Air bubbling can accelerate the procedure.

An alternative is to use 10% citric acid and a water spray, although the latter method has been reported to fail in some cases [Chu-Fei, H. Ho et al., 1994]. Detergents can be associated in the cleaning mixture and alternatives are also vinegar or ammonia [Martin, 1994]. In all cases, washing with a clear water bath or spray is recommended at the end of the procedure. Cleaning of the window of the photocells usually needs an additional mechanical brushing (softly, however, so as not to damage the window material). Calibration of the cell after cleaning the window is required.

5.6 TOXICITY AND FORMATION OF BY-PRODUCTS

At the UV doses applied for wastewater disinfection (with some exceptions), pre-existing potentially toxic compounds are not significantly removed. For synergistic technologies, see Chapter 4.

Formation of aldehydes has been observed both with low-pressure lamps [Awad et al., 1993] and medium-pressure lamps [Soroushian et al., 1994]. In summary, at irradiation doses of 1000 and 2000 J/m^2:

- Volatile and semivolatile compounds (EPA 8270) such as chloroform and other chlorinated by-products, 2-hexanone at parts per billion levels are removed by the general treatment without evidence of impact of UV.
- Carboxylic acids (acetic, formic, oxalic, haloacetic acids) at subparts per billion levels are unchanged with UV.
- Aldehydes (formaldehyde, acetaldehyde, glyoxal, m-glyoxal) are potentially formed at parts per billion levels.
- Alcohols (butanol, pentanol) are nondetectable in the UV-treated effluent.
- Propanol and substituted propanols—2-(2-hydroxypropoxy)-1-propanol, 1-(2-ethoxypropoxy)-2-propanol, 1-(2-methoxypropoxy)-1-propanol—at parts per billion levels appear unchanged at the preceding UV irradiation doses.

Toxicity to fish of UV-treated wastewater tested both in laboratory and at full scale did not show any additional toxicity vs. that of the effluent before UV treatment [Cairns and Conn, 1979; Oliver and Carey, 1976; Whitby et al., 1984].

5.7 PRELIMINARY CONCLUSIONS ON WASTEWATER DISINFECTION WITH ULTRAVIOLET

1. UV light technologies certainly are a valuable alternative for the disinfection of conventionally treated wastewater.
2. A wide choice of alternatives exists for lamp technologies and reactor designs.

3. Due to the variability of wastewater, an inventory of essential properties (at least bacterial counts in the effluent, TSS, and UVT, but also temperature, pH, etc.) is required to define design concepts.
4. Targets to be reached are also very variable as a function of local regulations. Therefore, targets must be clearly defined at a stage preceding the design.
5. When possible and for reaching particular targets, a pilot investigation is recommended.
6. How UV units can be installed in retrofitting of preexisting basins is described in a very documented way.
7. At conventional doses for disinfection, the formation of by-products is very marginal and no additional toxicity for fish life has been reported.
8. Elimination of preexisting potentially toxic compounds, in particular effluents (see Chapter 4), may need a point-of-use evaluation.

5.8   EXAMPLE

Figure 103 shows the UV installation at the Newcastle, Indiana wastewater treatment plant.

FIGURE 103 Plant at Newcastle, Indiana. This plant (designed by Berson) can treat 1570 m^3/h of a treated effluent with a transmittance of $T_{10\%}$ of 60. Two chambers are each equipped by 24 lamps mounted in the transverse mode. The equipment has been installed in preexisting buildings.

6 General Conclusions

- Both drinking water and process water of high quality can be disinfected by available ultraviolet (UV) technologies. The same holds for treated wastewaters.
- Precise guidelines apply to the design of equipment. Very often the information made available to the client remains somewhat empirical and merely commercial.
- By integrating present knowledge and experience, it is possible now to integrate adequate rules for design and methods for evaluation of performance. Depending on the case, tentative general rules are indicated in this text (Sections 2.6, 3.10, 4.5, and 5.7).
- The choice of a given technology and the performance of a given option of the UV lamps are determinants, depending on the expected result. Tailoring to measure is possible at present (see Chapter 2).
- The selection of the lamp is a determining issue that depends on any particular application. Constant progress is being made in the field (see Chapters 2 and 3).
- Appropriate design of reactors is an element for the success of the method, to be evaluated in each case (see Chapters 3 and 5).
- In all instances, essential data on the general quality of the water remain necessary, such as total suspended solids, transmittance of UV light, concentration of dissolved oxygen, turbidity, iron content, and general ionic balance.
- Fundamental principles of the application are at present thoroughly grounded and explained in this text.
- Cost parameters may be very case dependent. These are not commented on in this contribution and should be evaluated for each specific application.

Glossary*

absorption coefficient. *See* **Beer–Lambert law.**

Beer–Lambert law. Quantifies the absorption of a monochromatic wavelength by an homogeneous substance relating the incident radiant intensity (I_o) to the transmitted intensity (I) by $I = I_o \times 10^{-ACd}$, where A is the absorption coefficient; C, the concentration of the absorbing species, and d, the optical path length (usually in centimeters). The units for A are either in liters per mole-centimeter, or, for undefined compounds or for mixtures, in liters per centimeter (of liquid).

Alternatively, the law can be expressed basically on the Naperian log(ln) basis: $I = I_o\, e^{-ECd}$, in which E is the extinction coefficient, also in liters per mole-centimeter. Other symbols and units are found in the literature; however, the important point is to distinguish data expressed in publications either in the Log 10 base or in the Ln e base logarithms.

black body. A thermodynamic equilibrium concept that correlates temperature-heat transfer into the capacity of emission of light at given wavelengths. The maximum of emission is displaced to lower wavelengths when the temperature is increased.

Bunsen–Roscoe law. The ratio of reaction proportional to the absorbed dose. (The law generally applies to disinfection of drinking water; in photochemical processes, side effects must be considered.)

constants. *See* end of this glossary.

dose. Corresponds to the radiant power or radiant flux received per second by a unit surface. In this text, the dose is expressed in joules per square meter; in the literature data often are found also in milliwatt second per square centimeter.

Einstein law. The absorption of one single photon promoting a single photochemically induced change in the absorbing atom or molecule. The initial change in the molecule is the result of the absorption of one single photon.

Einstein (unit). The Einstein can be considered as a mole of photons (i.e., 6.022×10^{23} photons of the wavelength considered). For example, at 253.7 nm, 1 E is equal to 472 kJ, or 131 Wh (or $1J = 1\ W \cdot sec = 2.12\ E$).

energy. Energy is expressed in joule. The energy of a photon is given by $E = h\nu = hc/\lambda$; where h is the Planck constant (6.626×10^{-34} J sec) and c, the velocity of light (2.998×10^8 m/sec).

energy of a photon. *See* **energy.**

frequency. May be expressed in hertz (i.e., cycles per second; sec^{-1}), or wave number $1/\lambda$ (per meter or per centimeter).

---

* Note that units, terms, and symbols are as in the Système International (SI) system.

Grothius–Draper law. Only radiation (photon) that is absorbed capable of initiating a photochemical process.

intensity. Flux or fluence (i.e., power), incident on a surface of unit area; watt per square meter. (This is not to be confused with radiant intensity; *see also* **irradiance.**)

irradiance. *See* **intensity.** *Remark*: both intensity and irradiance are found indistinctly in the literature on water treatment. Irradiance is a nonspecific concept concerning wavelength, emission source, and distance from the source. Intensity (less useful for general lighting conditions) remains more specifically wavelength-related and involves more discrete (specific) source receptors (i.e., specific parts of DNA instead of general irradiance).

length. In meters (m) or centimeters (cm). Wavelengths usually are expressed in nanometers (nm) or micrometers (μm). (In the literature, one can still find units that are not in conformity with the SI nomenclature such as microns (μ), which equals micrometers; millimicrons (mμ), which equals nanometers, and Ångströms (Å), which equals 10^{-9} m.)

photometry. Measurement of light energy perceived by the human eye. Many photometric units exist, such as lux, lumen, candella, phots, etc. In UV parlance, these units are not used.

Planck constant. Proportionality constant between radiant energy and frequency of light $E = h\nu$, with $h = 6.626 \times 10^{-34}$ J sec.

Planck's theory. Electromagnetic radiation consists of discrete quanta (or photons) quantified by the energy of each photon as $h\nu$ (*see* **Planck constant**).

radiance. Flux (power) per unit solid angle per unit surface area (remote source) watt per square meter per steradian (W m^{-2} sr^{-1}).

radiant emittance. Flux per unit area received from a remote source: watt per square meter (W m^{-2}).

radiant energy. Radiant power multiplied by the irradiation time: watt times second (W × sec) or Joule (J).

radiant intensity. Flux (power) emitted by a source per unit solid angle: watt per steradian (W sr^{-1}).

radiant power or radiant flux. Emitted power by a light source, watt (W).

radiometry. Quantification of total radiant energy at all wavelengths emitted by a source.

radiometry (spectral). *See* **spectral radiometry.**

reciprocity law. *See* **Bunsen–Roscoe law.**

spectral radiometry. Quantification of radiant energy emitted at particular wavelengths or wavelength regions.

wavelength. *See* **length.**

Constants

Constant	Symbol	Unit	Value
Avogadro constant	N_A	mol^{-1}	6.02×10^{23}
Boltzman constant	k	$J\ mol^{-1}\ K^{-1}$	1.38×10^{-23}
Electron electrical charge	e	C (coulomb)	1.6×10^{-19}
Faraday constant	F	$C\ mol^{-1}$	9.65×10^4
Gas constant	R	$J\ mol^{-1}\ K^{-1}$	8.315
Gravity (acceleration)	g	$m\ sec^{-2}$	9.81
Joule/cal	J		4.184
Molar gas volume at NTP[a]	—	$m^3\ mol^{-1}$	2.2414×10^{-2}
Planck constant	h	J sec	6.626×10^{-34}
	$h/2\pi$	J sec	1.055×10^{-34}
Speed of sound (NTP[a])	C_s	$m\ sec^{-1}$	331.45
Temperature (absolute)	K		273.15°C
Velocity of light	c	$m\ sec^{-1}$	3×10^8 (vacuum)
Einstein	E (= 1 mol of photons)		

[a] NTP, normal temperature and pressure.

References

Acero, J.L. and von Gunten, U., in Proc. IOA-EA3G Symposium, Poitiers, France, International Ozone Association, Paris, France, 1998.

Aicher, J.O. and Lemmers, E., *Illum. Eng.*, 52, 579, 1957.

Akkad, F., Pape, E., and Weigand, F., *IHLE Berichte*, Düsseldorf, Germany, 1986.

Aklag, M.S., Schumann, H.P., and von Sonntag, Cl., *Environ. Sci. Technol.*, 24, 379, 1990.

Al Ekabi, H., *AOTs,* Science and Technology Integration, University of Western Ontario Research Park, London, Ontario, Canada, 1994.

Allen, A.O., *J. Phys. Colloidal Chem.*, 52, 479, 1948.

Anon., *Eng. News,* 66, 686, 1911.

Anon., Sterilization of polluted water by ultra-violet rays at Marseille (France), *Eng. News,* 64, 633, 1910.

Appleton, A.R. et al., communication, Montgomery Watson Engineers, Walnut Creek, CA, 1994.

AQUA—J. Water Supply, 16(2), 1992.

ATV, St. Augustin, (now Hennef, Germany), 1993.

Awad, J., Gerba, C., and Magnuson, G., in Proc. WEF Speciality Conference, Whippany NJ, Water Environment Federation, Alexandria, VA, 1993.

AWWA, *J. Am. Water Works Assoc.*, 60, 1317, 1968.

(Austria) Österreichisches Normungsinstitut (Austrian Standard Institute), Vienna, Austria, ÖNORM M 5873 (1996); ÖNORM M 5873-1, ed., 2001-03-01, 2001.

AWWA, Recommendations for raw water quality, *J. Am. Water Works Assoc.*, 60, 1317, 1968.

Bandemer, Th. and Thiemann, W., in Proc. IOA Symposium, Amsterdam, the Netherlands, International Ozone Association, Paris, France, 1986, p. C.4.1.

Barich, J.T. and Zeff, J.D., in Proc. 83rd Annual Meeting and Exhibition of Air & Water Management Association, Vol. 2, AWMA, Pittsburgh, PA, 1990, p. 90–245.

Baron, J. et al., *Technol. Sci. Méthodes-AGHTM,* 1999.

Baxendale, J. and Wilson, J., *Trans. Faraday Soc.*, 53, 344, 1957.

Bayliss, S. and Bucat, R., *Aust. J. Chem.*, 28, 1865, 1975.

Becker, H.G.O., *Einführung in die Photochemie,* 2nd ed., George Thieme Verlag, Stuttgart, Germany, 1983.

Benitez, F.J., Beltrán-Heredia, J., and Gonzalez, T., *Ozone: Sci. Eng.*, 16, 213, 1994.

Bernardin, F.E., in Proc. 84th Annual Meeting and Exhibition of the Air and Water Management Association, Vancouver, BC, Canada; as cited in Air and Water Management Association, Pittsburgh, PA, 1991, pp. 11, 91-24-1.

Bernardt, H. et al., in Proc. IOA Conference, Amsterdam, the Netherlands, 1986, International Ozone Association, Paris, France, 1986, p. 85; see also Masschelein, W.J., 1986.

Bernhardt, H. et al., *Wasser-Abwasser,* 133, 632, 1992.

Beyerle-Pfnür et al., *Toxicol. Environ. Chem.*, 20–21, 129, 1989.

Bischof, H.M.A. thesis, Technical University of Munich, Germany, 1994.

Bission, J.W. and Cabelli, V.W., *J. Water Pollut. Control Fed.*, 52, 241, 1980.

Blatchley, E.R. and Xie, Y., *Water Environ. Res.*, 67, 475, 1995.

Blatchley, E.R. et al., communication, School of Civil Engineering, Environmental and Hydrological Engineering, Purdue University, West Lafayette, IN, 1994.

Blomberg, J., Ericksson, U., and Nordwall, I., UV Disinfection and Formation of Chlorinated By-Products in Presence of Chloramines, in Proc. Intil. Conf. on Applications of Ozone and also on UV and Related Ozone Technologies (in Conjunction with IUVA) at Wasser Berlin 2000, International Ozone Association, Paris, France, 2000, pp. 329– 344.

Bolton, J.R., *Eur. Photochem. Assoc. Newsl.*, 43, 40, 1991.

Bors, W. et al., *Photochem. Photobiol.*, 28, 629, 1978.

Bossuroy A. and Masschelein, W.J., unpublished observations, 1987.

Bott, W., *Zentralbl. Bakteriol. Hyg. Abt. Orig. B*, 178, 263, 1983.

Bourgine, F.P. and Chapman, J.E., in Proc. IOA Conference, Amsterdam, the Netherlands, 1986, International Ozone Association, Paris, France, 1996, p. II.a; see also Masschelein, W.J., (1996a,b).

Braun, A., Maurette, M.-Th., and Oliveros, E., Technologie Photochimique, Ed. Presses Poly-techniques Romandes Diffusion par Lavoisier Technique et Documentation, Paris, France, 1986.

Braunstein, J.F. et al., *WEF, Digest,* 1994, p. 335.

Bruce, J., *Gen. Physiol.,* 41, 693, 1958.

Buck, R.P., *Anal. Chem.,* 26, 1240, 1954.

Bukhari, Z., Hargy, T.M., Bolton, J.R., Dussert, B., and Clancy, J.L., Medium-pressure UV light for oocyst inactivation, *J. Am. Water Works Assoc,* 91(3), 86–94, 1999.

Cabaj, A., Sommer, R., and Haider, Th., in Proc. IOA Conference Wasser Berlin—2000, International Ozone Association, Paris, France, 2000, pp. 297–311.

Cairns, V.W. and Conn, K., Canada-Ontario Agreement of Great Lakes Water Quality, Research report 92, 1979.

Calvert, J.G. and Pitts, J.N., *Photochemistry,* John Wiley & Sons, New York, 1966.

Carey, J.H., Lawrence, J., and Tosine, H.M., *Bull. Environ. Toxicol.,* 16, 697, 1976.

Carnimeo, D. et al., *Water Sci. Technol.,* 30, 125, 1994.

Carter, S.R. et al., U.S. Patent 5, 043, 080, 1991.

Castrantas, H.M. and Gibilisco, R.D., *ACS Symp. Ser.,* 422, 77, 1990.

Cayless, M.A., *Br. J. Appl. Phys.,* 11, 492, 1960.

Chameides, W.L. and Walker, J.G.C., *J. Geophys. Res.,* 32, 89, 1997.

Chu-Fei, H. Ho, et al., communication environmental engineering, San Francisco, CA, 1994.

Clancy, J.L., Bukhari, Z., Hargy, T.M., Bolton, J.R., Dussert, B., and Marshall, M.M., Using UV to inactivate *Cryptosporidium, J. Am. Water Works Assoc.,* 92(9), 97–104, 2000.

Clancy, J.L. and Hargy, T.M., Ultraviolet light inactivation of *Cryptosporidium, Chem. Technol.,* July–Aug., 5–7, 2001.

Clancy, J.L., Hargy, T.M., Marshall, M.M., and Dyksen, J.E., UV light inactivation of *Cryptospo-ridium* oocysts, *J. Am. Water Works Assoc.,* 90(9), 92–102, 1998.

Clemence, W., UV at Marseille, *Engineering,* 91, 106, 142, 1911.

Cortelyou, J.R. et al., *Appl. Microbiol.,* 2, 227, 1954.

Denis, M., Minon, G., and Masschelein, W.J., *Ozone: Sci. Eng.,* 14, 215, 1992.

Deutschland: Vorschriften für Klassification und Bau von Strählern; Seeschiffe, Band II, Kap. (Aug. 1973).

DIN (German Normalization Institute), Standard 5031-10, 1996.

Dodin, A. et al., *Bull. Acad. Nat. Médecine (France),* 155, 44, 1971.

Dohan, J.M. and Masschelein, W.J., *Ozone: Sci. Eng.,* 9, 315, 1987.

Downes, A. and Blunt, T.P., *Proc. R. Soc.,* 26, 488, 1877.

Dulin, D., Drossman, H., and Mill, T., *Environ. Sci. Technol.,* 20, 20, 1986.

Dussert, B.W., Bircher, K.G., and Stevens, R.D., in Proc. IOA Conference, Amsterdam, the Netherlands, 1996, International Ozone Association, Paris, France, 1996, p. I.IV.a.

DVGW Arbeitsblatt, 1997, W-29–4.

Dwyer, R.J. and Oldenburg, O., *J. Chem. Phys.*, 12, 351, 1944.

Eaton, A., *J. Am. Water Works Assoc.*, 87, 86, 1995.

Egberts, G., in Proc. IOA Conference Wasser Berlin—1989, International Ozone Association, Paris, France, 1989, pp. II-1–10.

Eliasson, B. and Kogelschatz, U., in Proc. IOA Conference Wasser Berlin—1989, International Ozone Association, Paris, France, 1989, p. IV-6–4.

Elenbaas, W., *The High Pressure Mercury Discharge*, North-Holland, Amsterdam, 1951.

Ellis, C. and Wells, A.A., *The Chemical Action of Ultraviolet Rays*, Reinhold, New York, 1941.

EPRI, Electric Power Research Institute, Ultraviolet Light for Water and Wastewater, 1995.

Fair, G.M., *J. Am. Water Works Assoc.*, 7, 325, 1920.

Fassler, D. and Mehl, L., *Wiss. Z. Univ. Jena; Naturwiss. Reihe*, 20, 137, 1971.

FIGAWA, Technische Mitteilungen, Cologne, Germany, 1987.

Finch, G. and Belosevic, M., communication at U.S. EPA Workshop on NV for Potable Water Applications, Arlington, VA, April 28–29, 1999.

Fletcher, D.B., *WaterWorld News*, 3, May–June, 1987.

Francis, P.D., in Proc. IOA Conference, Amsterdam, the Netherlands, 1986, International Ozone Association, Paris, France, 1986, pp. C.3.1–17.

Francis, P.D., Electricity Council Centre, Chester, U.K., 1988, report M 2058.

Frischerz, H., et al., *Water Supply*, 4, 167, 1986.

Geesey, G.G. and Costerson, J.W., *Can. J. Microbiol.*, 25, 1058, 1984.

Gellert, B. and Kogelschatz, U., *Appl. Phys. B*, 12, 14, 1991.

Gelzhäuser, P., *Disinfection von Trinkwasser durch UV-Bestrahlung*, Expert Verlag Ehningen, Ehningen, Germany, 1985.

Germany: Seeschiffen, Vol. 2, Chap. 4, 1973.

Germany: Bayrisches Landesamt für Wasserwirtschaft, 1.-7.3, 1982.

Glaze, W.H., An overview of advanced oxidation processes: current status and kinetic models, in *Chemical Oxidation, Technologies for the Nineties*, Vol. 3, Technomic Press, Lancaster, PA, 1994, pp. 1–11.

Glaze, W.H. and Kang, J.W., in Proc. Symposium on Advanced Oxidation Processes and Treatment of Contaminated Water and Air, Burlington, Ontario, Canada, 1990.

Glaze, W.H., Kang, J.W., and Chapin, D.H., *Ozone: Sci. Eng.*, 9, 335, 1987.

Goldstein, E., *Chem. Ber.*, 36, 3042, 1903.

Groocock, N.H., *J. Inst. Water Eng. Scientists*, 38, 163, 1984.

Guillerme, J., Collection Que Sais-Je, Presses Universitaires de France, 1974.

Guittonneau, S. et al., *Environ. Technol. Lett.*, 9, 1115, 1998.

Guittonneau, S. et al., *Ozone: Sci. Eng.*, 12, 73, 1990.

Guittonneau, S. et al., *Rev. Fr. Sci. Eau*, 1, 35, 1988.

Gurol, M.D. and Vastistas, R., *Water Res.*, 21, 895, 1987.

Haas, C.N. et al., U.S. EPA Design Manual, U.S. EPA Report No. 625/1-86/021, 1986.

Hargy, T.M., Clancy, J.L., Bukhari, Z., and Marshall, M.M., Shedding UV light on the *Cryptosporidium* threat, *J. Environ. Health*, July–Aug. 19–22, 2000.

Harm, W., *Biological Effects of Ultraviolet Radiation*, Cambridge University Press, Cambridge, U.K., 1980.

Harris, G.D., *Water Res.*, 21, 687, 1987.

Harris, L. and Kaminsky, J., *J. Am. Chem. Soc.*, 57, 1154, 1935.

Hashem, T.M. et al., in Proc. IOA Conference, Amsterdam, the Netherlands, 1986, International Ozone Association, Paris, France, 1996, pp. II.V.A.

Hatchard, G.C. and Parker, C.A., *Proc. R. Soc. A*, 235, 518, 1956.

Hautniemi, M. et al., *Ozone: Sci. Eng.*, 20, 259, 1997.

Havelaar, A.H. and Hogeboom, W.M., *J. Appl. Bacteriol.,* 56, 439, 1984.

Havelaar, A.H. et al., Chapter D-3; see also Masschelein, W. J., IOA Conference, Amsterdam, 1986, International Ozone Association, Paris, France, 1986.

Heath, M.S., private communication, Montgomery-Watson Engineers, at EPA Meeting on UV for Potable Water Applications, Arlington, VA, April 1999.

Himebaugh, W.S. and Zeff, J.D., in *Proc. Annu. Meet. Air Waste Manage. Assoc.,* 11, 91/24.3, 1991.

Ho, P.C., *Environ. Sci. Technol.,* 20, 260, 1986.

Hoigné, J. and Bader, H., *Vom Wasser,* 48, 283, 9, 1977.

Hoigné, J., *Handbook of Environmental Chemistry,* Vol. Part C, Springer-Verlag, Heidelberg, Germany, 1998, p. 83.

Hölzli, J.P., private communication, Hölzli Gesellschaft, Seewalchen, Austria, 1992.

Huff, C.B. et al., *Public Health Rep.,* 337, 1965.

Hunter, G.L. et al., *Water Environ. Technol.,* 41, 1998.

IUVA, International Ultraviolet Association, Proc. First Intl. Congress on UV Technologies, Washington, D.C., June 14–16, 2001.

Jackson, G.F., *Eur. Water Pollut. Control,* 18, 1994.

Jacob, S.M. and Dranoff, J.S., *J. Am. Inst. Chem. Eng.,* 16, 359, 1970.

Jagger, J., *Introduction to Research in Ultraviolet Photobiology,* Prentice Hall, New York, 1967.

Jepson, J.D., *Proc. Water Treat. Eng.,* 175, 1973.

J. Water Supply—AQUA, 16(2), 1992.

Kalisvaart, B.F., Microbiological Effects of Berson Multiwave UV Lamps, Berson UV Techniek, Neunen, the Netherlands, 1999.

Kalisvaart, B.F., Berson Document: Photoelectrical Effects of Berson Multiwave Lamps to prevent Microbial Recovery, Berson UV Techniek, Neunen, the Netherlands, 2000.

Kalisvaart, B.F., Berson Company, private communication, 2001.

Kawabata, T. and Harada, T., *J. Illum. Soc.,* 36, 89, 1959.

Kelner, A., *J. Gen. Physiol.,* 34, 835, 1951.

Kiefer, J., *Ultraviolette Strahlen,* Ed., Walter de Gruyter, Berlin, Germany, 1977.

Köppke, K.E. and von Hagel, G., *Wasser-Abwasser,* 132, 1990; 313, 1991.

Kornfeld, G., *Z. Phys. Chem. B,* 29, 205, 1935.

Kraljic, I. and Moshnsi, S. El., *Photochem. Photobiol.,* 28, 577, 1978.

Kruithof, J.C. et al., in Proc. IOA Conference, Amsterdam, the Netherlands, 1996, International Ozone Association, Paris, France, 1996, pp. IV.V.a., see also Masschelein, W.J., 1996.

Kruithof, J.C. and Kamp, P.C., in Proc. IOA Conference Wasser Berlin—2000, International Ozone Association, Paris, France, 2000, pp. 437–464, 631–660.

Ku, Y. and Hu, S.C., *Environ. Prog.,* 9, 218, 1990.

Kwan, A. et al., communication, CH2M HILL Engineering, Calgary, Alberta, Canada, 1994.

Laforge, P., Fransolet, G., and Masschelein, W.J., *Rev. Fr. Sci. Eau,* 1, 255, 1982.

Lafrenz, R.A., Workshop U.S. EPA, Arlington, VA, Apr. 28–29, 1999.

Lafrey, (copy at), EPA meeting on UV for Potable Water Applications, Arlington, VA, April 1999.

Legan, R.W., *Chem. Eng.,* 25, 95, 1982.

Legrini, O., Oliveros, E., and Braun, A.M., *Chem. Rev.,* 93, 671, 1993.

Leighton, W.G. and Forbes, G.S., *J. Am. Chem. Soc.,* 52, 3139, 1930.

Leitzke, O. and Friedrich, M., in Proc. IOA Conference Moscow, Russia, 1998, International Ozone Association, Paris, France, 1996, p. 567.

Leitzke, O., in Proceedings of IOA Conference at Wasser Berlin—1993, (Paris, France: International Ozone Association, 1993, p. IV.6.1; see also Brtko-Juray, and O. Leitzke, in Proc. IOA Conference at Wasser Berlin—1993, International Ozone Association, Paris, France, 1993, p. III.1.1.

Lenard, P., *Ann. Phys. (Leipzig)*, 1, 486, 1900.

Leuker, G. and Dittmar, A., BBR. Germany, 54, 1992.

Leuker, G. and Hingst, V., *Zentralbl. Hyg.*, 193, 237, 1992.

Linden, K.G., Workshop U.S. EPA, Arlington, VA, April 28–29, 1999.

Linden, K.G. and Darby, J.L., *J. Environ. Eng.*, 123, 1142, 1997.

Lindenauer, K.G. and Darby, J.L., *Water Res.*, 28, 805, 1994.

Lipczynska-Kochany, E. and Bolton, J.R., *Environ. Sci. Technol.*, 26, 259, 1992.

Lipczynska-Kochany, E., Degradation of aromatic pollutants by means of the advanced oxidation processes in a homogeneous phase: photolysis in the presence of hydrogen peroxide versus the fenton reaction, in *Chemical Oxidation, Technologies for the Nineties*, Vol. 3, Technomic Press, Lancaster, PA, 1994, pp. 12–27.

Lissi, E. and Heicklen, J., *J. Photochem*, 1, 39, 1973.

Lodge, F.J., reported in *WEF Digest*, p. 437, 1994.

Lowke, J.J. and Zollweg, R.H., *J. Illum. Eng. Soc.*, 4, 253; *J. Appl. Phys.*, 46, 650, 1975.

Maier, A. et al., *Water Sci. Technol.*, 31, 141, 1995.

Malley, J., personal communication, University of New Hampshire, Durham, NH, 1999.

Mark, G. et al., *Aqua*, 39, 309, 1990; *J. Photochem. Photobiol. A—Chem.*, 55, 1990.

Martin, J.S., reported in *WEF Digest*, p. 461, 1994.

Martiny, H. et al., *Zentralbl. Bakteriol.*, 185, 350, 1988.

Masschelein, W., *TSM Eau*, 61, 95, 1966.

Masschelein, W.J., *TSM Eau*, 77, 177, 1977.

Masschelein, W.J., Coordinated Proceedings of IOA Conferences, Amsterdam, International Ozone Association, Paris, France, 1986 and 1996.

Masschelein, W.J., *Unit Processes in Drinking Water Treatment*, Marcel Dekker, New York, 1992.

Masschelein, W.J., in *Proc. CEFIC Symposium, Rome, Italy, 1996a*, p. 153.

Masschelein, W.J., *Processus Unitaires dans le Traitement de l'Eau Potable*, CEBEDOC, Liège, Belgium, 1996b.

Masschelein, W.J., in Proc. 14th Ozone World Congress, Dearborn, MI, International Ozone Association, Pan American Group, Norwalk, CT, 1999.

Masschelein, W.J., *Wasser*, in Proc. IOA Conference, Berlin, 2000, International Ozone Association, Paris, France.

Masschelein, W.J., Debacker, E., and Chebak, S., *Rev. Fr. Sci. Eau*, 2, 29, 1989.

Masschelein, W.J., Fransolet, G., and Debacker, E., *Eau Québec*, 13, 289; 14, 41, 1981.

Masten, S.J. and Butler, J.N., *Ozone: Sci. Eng.*, 8(4), 339–353, 1986.

Mathews, R.W., *J. Chem. Soc. Faraday Trans.*, 85, 1291, 1989.

Mazoit, L.P. et al., *Trib. Cebedeau*, 344, 21, 1975.

McGregor, F.R., in Proc. IOA Conference, Amsterdam, the Netherlands, 1986, International Ozone Association, Paris, France, 1986, p. B.5.1.

Mechsner, K.I. and Fleischmann, Th., *GWA*, 72, 807, 1992.

Meulemans, C.C.E., in Proc. IOA Conference, Amsterdam, the Netherlands, 1986, International Ozone Association, Paris, France, 1986, chap. B-1; see also Masschelein, W.J., 1986.

Milano, J.C., Bernat-Escallien, C., and Vernet, J.L., *Water Res.*, 24, 557, 1990.

Murov, S.L., Transmission of light filters, in *Handbook of Photochemistry*, Marcel Dekker, New York, 1973, pp. 97–103.

Narkis, N. et al. in *Water Chlorination,* Vol. 6, Jolley, R.L., Ed., Lewis Publishers, Chelsea, MI, 1987, chap. 73.

Nicole, I. et al., *Environ. Technol.,* 12, 21, 1991.

NWRI, *Ultraviolet Disinfection Guidelines for Drinking Water and Water Reuse,* National Water Research Institute, Fountain Valley, CA, Dec. 2000.

Okatomo, K. et al., *Bull. Chem. Soc. Jpn.,* 58, 2015; 2023, 1985.

Oliver, B.G. and Carey, J.H., *J. Water Pollut. Control Fed.,* 48, 2619, 1976.

Österreichisches Normungsinstitut (Austrian Standards Institute, Vienna, Austria), ÖNORM M 5873, 1996; ÖNORM M 5873-1, ed. 2001-03-01, 2001.

Paillard, H., thèse, Université de Poitiers, Poitiers, France, 1996.

Parker, J.A. and Darby, J.L., reported in *WEF Digest,* p. 469, 1994.

Pellizetti, E. et al., *Chim. Ind.,* 67, 623, 1985.

Perkins, R.G. and Welch, H., *J. Am. Water Works Assoc.,* 22, 959, 1930.

Peterson, D., Watson, D., and Winterlin, W., *Bull. Environ. Contam. Toxicol.,* 44, 744, 1990.

Pettinger, K.H., Entwicklung und Untersuchung eines Verfahrens zum Atrazineabbau in Trinkwasser mittels UV-aktiviertem Wasserstoffperoxid, thesis, Technical University of Munich, Germany, 1992.

Peyton, G.R., Oxidative treatment methods for removal of organic compounds from drinking water supplies, in *Significance and Treatment of Volatile Organic Compounds in Water Supplies,* Lewis Publishers, Chelsea, MI, 1990, pp. 313–362.

Peyton, G.R. and Gee, C.S., in *Advances in Chemistry Series 219,* American Chemical Society, Washington, D.C., 1989, p. 639.

Peyton, G.R. and Glaze, W.H., *Environ. Sci. Technol.,* 22, 761, 1988.

Peyton, G.R. and Glaze, W.H., Photochemistry of environmental systems, in *Advances in Chemistry Series 327,* American Chemical Society, Washington, D.C., 1986, p. 76.

Phillips, R., *Sources and Applications of Ultraviolet Radiation,* Academic Press, 1983.

Prado, J. et al., *Ozone: Sci. Eng.,* 16, 235, 1994.

Prat, C., Vincente, M., and Esplugas, S., *Ind. Eng. Chem. Res. Ser.,* 29, 349, 1990.

Qualls, R.G. and Johnson, J.D., *Appl. Environ. Microbiol.,* 45, 872, 1983.

Rabinowitch, E., *Photosynthesis,* Interscience, New York, 1945.

Rahn, R.O., *Photochem. Photobiol.,* 66, 450, 1997.

Ridgway, H.F. and Olson, B.H., *Appl. Environ. Microbiol.,* 41, 274, 1981; 44, 972, 1982.

Roper, M.M. and Marshall, K.C., *Microbial Ecol.,* 4, 279, 1978.

Rudolph, K.U., Böttcher, J., and Nelle, Th., *GWF; Wasser-Abwasser,* 135, 529, 1994.

Sadoski, T.T. and Roche, W.J., *J. Illum. Eng. Soc.,* 5, 143, 1976.

Sagawura, T., Funayama, H., and Nakano, K., presentation at Am. Inst. Chem. Engrs. Symposium on Photochemical Reaction Engineering, Institute of Chemical Engineers, New York, NY, 1984.

Schäfer, J., British Patent 1,552,334, 1979.

Scheible, O.K., Casey, M.C., and Fondran, A., NITS Publication 86-145182, National Technical Information Service, Springfield, VA, 1985.

Schöller, F. and Ollram, F., Water Supply, Special Session 19, 1989, p. 13.

Schumb, W. and Satterfield, C., *Hydrogen Peroxide,* Reinhold, New York, 1955.

Severin, B.F. et al., *J. Water Pollut. Control Fed.,* 56, 164; 881, 1984.

Sharpatyi, V. and Kraljic, I., *Photochem. Photobiol.,* 28, 587, 1978.

Shi, H., Wang, Z., and Zhang, C., in Proc. IOA Symposium, Amsterdam, International Ozone Association, Paris, France, 1986, p. C.1.1–10.

Shuali, U. et al., *J. Phys. Chem.,* 73, 3445, 1969.

Sierka, R.A. and Amy, G.L., *Ozone: Sci. Eng.,* 7, 47, 1985.

Simmons, M.S. and Zepp, G.R., *Water Res.,* 20, 899, 1986.

Smith, A.T., *Eng. News Rec.*, 79, 1021, 1917.

Smithells, C.J., *Metals Reference Book*, 5th ed., Butterworths, London, 1976.

Sommer, R. et al., *Water Sci. Technol.*, 35(11/12), 113, 1997.

Soroushian, F. et al., communications, CH2M HILL, Santa Ana, CA, 1994.

South African General and Special Standards, Act 96 of 18 May 1984, 9225—Regulation 991.

Spencer, R.R., *J. Am. Water Works Assoc.*, 4, 172, 1917.

Staehelin, J., and Hoigné, J., *Environ. Sci. Technol.*, 19, 1206, 1985.

Stryer, L., *Biochemistry*, W.H. Freeman, San Francisco, CA, 1975.

Sundstrom, D.W. and Klei, H.E., NTIS Publ. Nr. PB 87-149357, National Technical Information Service, Springfield, VA, 1986.

Sundstrom, D.W. et al., *Hazard Waste—Hazard Mater.*, 3, 101, 1986.

Sundstrom, D.W., Weir, B.A., and Reding, K.A., in *ACS Symp. Ser.*, 422, 67, 1989.

Symons, J.M., Prengle, H.W., and Belhateche, D., in Proc. Am. Water Works Assoc. Conf., 422, part 1, p. 67; part 2, p. 895, 1989.

Takahashi, N., Ozonation of several organic compounds having low molecular weight under ultraviolet irradiation, *Ozone: Sci. Eng.*, 12(1), 1–17, 1990.

Taube, H. and Bray, W.H., *J. Am. Chem. Soc.*, 62, 3357, 1940.

Te Kippe, T. et al., reported in *WEF Digest*, p. 511, 1994.

Thampi, M.V. and Sorber, C.A., *Water Res.*, 21, 765, 1987.

Torota, *Microbiology; An Introduction*, 5th ed., Benjamin Cummings, Menlo Park, CA, 1995.

Toy, M.S., Carter, M.K., and Pasell, T.O., *Environ. Sci. Technol.*, 11, 837, 1990.

Trapido, M. et al., *Ozone: Sci. Eng.*, 19, 75, 1997.

Tree, J.A., Adams, R.R., and Lees, D.N., *Water Sci. Technol.*, 35, 227, 1997.

United Kingdom, Regulation 29(6) of the Merchant Shipping Regulations of the Marine Surveyors of the Department of Trade, 1973.

United States, *Handbook on Sanitation of Vessel Construction*, U.S. Government Printing Office, Department of Health, Education and Welfare, Washington, D.C., pp. 17–18.

Uri, N., Inorganic free radical reactions in solution, *Chem. Rev.*, 50, 375, 1952.

U.S. Department of Commerce, Investigation into the Chemistry of the UV-Ozone Purification Process, National Technology Information Service, 1979, Springfield, VA, report PB-296 485.

U.S. EPA, Design Manual for Municipal Wastewater Disinfection, report EPA/625/1-86/021 U.S. Environmental Protection Agency, Washington, D.C., 1986.

U.S. EPA, Workshop on UV Disinfection of Drinking Water, Arlington, VA, 1999.

U.S. EPA, National primary drinking water regulations: ground water rule; proposed rules, *Fed. Reg.*, 65(91), 30193, 2000.

von Recklinghausen, M., *J. Am. Water Works Assoc.*, 1, 565, 1914.

Walden, F.H. and Powell, S.T., in Proc. Am. Water Works Assoc. Conference, 1911, p. 341.

Waymouth, J.F., *Electric Discharge Lamps*, MIT Press, Cambridge, MA, 1971.

Wayne, R.P., *Faraday Discuss. Chem. Soc.*, 53, 172, 1972.

WEF Digest, Water Environmental Federation, Alexandria, VA; Available also at European Office, The Hague, the Netherlands, 1994.

Weir, B.A. and Sundstrom, D.W., in Proc. Am. Inst. Chem. Engrs. Natl. Meeting, San Francisco, CA, 1989.

Weir, B.A., Sundstrom, D.W., and Klei, H.E., *Hazard Waste—Hazard Mater.*, 4, 165, 1987.

Weiss, J., *Symposium on Mechanisms of Electron Transfer Reactions*, Société de Chimie Physique, Paris, France, 1951.

Whitby, G.E. et al., *J. Water Pollut. Control Fed.*, 57, 844, 1984.

Winter, Br., dissertation, Technical University of Munich, Germany, 1993.

Wright, H., Cairns, W.L., and Sakamoto, G., communication, Trojan Technologies, London, Ontario, Canada, 1999.

Xu, S., Zhou, H., Wei, X., and Lu, J., *Ozone: Sci. Eng.,* 11, 281, 1989.

Yost, K.W., in Proc. Intl. Waste Conf. Water Pollut. Control Fed., 1989, p. 441.

Yue, P.L. and Legrini, O., Am. Inst. Chem. Engrs. Natl. Meeting, San Francisco, CA, 1989.

Zeff, J.D. and Leitis, E., U.S. Patent 4,792,407, 1988; U.S. Patent 4,849,114, 1989.

Zukova et al., *J. Water Pollut. Control Fed.,* 58, 199, 1986.

Index